THE
DOSE
EFFECT

About the Author

TJ Power is a neuroscientist, international speaker, and co-founder of Neurify, a mental health and performance training company. During his BSc and MSc studies TJ focused on psychology, health science, and neuroscience and identified a significant gap in mental health support within educational and corporate environments. As a result, he conceptualized The DOSE Effect, a science-backed formula designed to help people focus on taking action towards their health goals by harnessing four key chemicals: Dopamine, Oxytocin, Serotonin, and Endorphins. This idea led to the launch of Neurify and the research center, The DOSE Lab, where the foundation of his book research is hosted. Throughout the early years of Neurify, TJ delivered DOSE Live extensively, personally training over 50,000 people within two years, achieving remarkable results.

TJ's work has since gained significant recognition worldwide for its innovative approach to mental health, particularly in addressing the challenges of modern, digitally driven lives. He has worked with organizations such as Amazon, Coca-Cola, the National Health Service (NHS), and Oxford University. With over 500 keynotes delivered across the UK and internationally, TJ consistently demonstrates his commitment to enhancing society's mental health through accessible and scientifically backed teachings.

To find out how you can experience DOSE Live with TJ himself, visit **www.tjpower.co.uk**

 @tjpower

THE DOSE EFFECT

TJ POWER

HQ

HQ
An imprint of HarperCollins*Publishers* Ltd
1 London Bridge Street
London SE1 9GF

www.harpercollins.co.uk

HarperCollins*Publishers*
Macken House, 39/40 Mayor Street Upper
Dublin 1, Ireland, D01 C9W8

This edition 2025

7

First published in Great Britain by
HQ, an imprint of HarperCollins*Publishers* Ltd 2025

Copyright © TJ Power 2025

TJ Power asserts the moral right to be
identified as the author of this work.
A catalogue record for this book is
available from the British Library.

HB ISBN: 978-0-00-866733-7
TPB ISBN: 978-0-00-866732-0

This book contains FSC™ certified paper and other controlled sources
to ensure responsible forest management.

For more information visit: www.harpercollins.co.uk/green

All illustrations by Chris Robinson

Design by Studio Nic&Lou
www.nicandlou.com

Printed and bound in the UK using 100% Renewable
Electricity at CPI Group (UK) Ltd

All rights reserved. No part of this publication may be reproduced, stored in a retrieval system, or transmitted, in any form or by any means, electronic, mechanical, photocopying, recording or otherwise, without the prior written permission of the publishers.

Without limiting the author's and publisher's exclusive rights, any unauthorised use of this publication to train generative artificial intelligence (AI) technologies is expressly prohibited. HarperCollins also exercise their rights under Article 4(3) of the Digital Single Market Directive 2019/790 and expressly reserve this publication from the text and data mining exception.

This book contains advice and information relating to health care. It should be used to supplement rather than replace the advice of your doctor or another trained health professional. If you know or suspect you have a health problem, it is always recommended that you seek your physician's advice before embarking on any medical program or treatment. All efforts have been made to assure the accuracy of the information contained in this book as of the date of publication. The publisher and the author disclaim liability for any medical outcomes that may occur as a result of applying the methods suggested in this book and any actions are taken entirely at the reader's own risk.

I dedicate this book to you and the transformational health journey you are about to experience. Thank you for being here.

Contents

WELCOME TO *THE DOSE EFFECT*	**8**
INTRODUCTION TO TJ POWER	**16**
PART 1: **DOPAMINE**	**20**
Chapter 1: Flow State	40
Chapter 2: Discipline	54
Chapter 3: Phone Fasting	62
Chapter 4: Cold Water	76
Chapter 5: My Pursuit	86
PART 2: **OXYTOCIN**	**98**
Chapter 6: Contribution	108
Chapter 7: Touch	116
Chapter 8: Social Life	126
Chapter 9: Gratitude	142
Chapter 10: Achievements	152

| **PART 3:** | **SEROTONIN** | **162** |

    Chapter 11: Nature — 172
    Chapter 12: Sunlight — 180
    Chapter 13: Gut Health — 188
    Chapter 14: Underthinking — 204
    Chapter 15: Deep Sleep — 214

| **PART 4:** | **ENDORPHINS** | **226** |

    Chapter 16: Exercise — 234
    Chapter 17: Heat — 248
    Chapter 18: Music — 256
    Chapter 19: Laughter — 264
    Chapter 20: Stretching — 270

CONCLUSION — **282**

YOUR DOSE ACTIONS — **292**

DOSE STACKING — **296**

THE DOSE REVOLUTION — **298**

INDEX — **300**

ACKNOWLEDGEMENTS — **304**

REFERENCES — **304**

Welcome to *The DOSE Effect*

In this book, I am going to tell you a scientific story. A story of an alternative route forward for us in our modern world. One in which we embrace the ever-changing technological experience that is coming, as well as reconnecting us to the path that brought us here. I am going to give you a new way to look at how you engage with technology, your life, your work, your relationships, and your health. A way in which discipline and motivation arise naturally and with ease. A way in which your life becomes beautiful, an energizing experience that is your new reality, every day.

DOSE is an acronym for the four key chemicals that live within your brain and body. These are **Dopamine, Oxytocin, Serotonin,** and **Endorphins**. These chemicals have evolved within us throughout our 300,000 years of human development.[1] These chemicals are our friends. They are here to guide us towards the best experience of life we can possibly have.

Our goal throughout *The DOSE Effect* is simply to learn how to listen to them. Once you understand how they impact your feelings, you will then learn to be guided by them. Each of them has a very specific function for us.

UNDERSTANDING
The Chemicals

D
Dopamine
THE MOTIVATIONAL CHEMICAL

Built through hard work and effort, dopamine creates your drive and motivation to pursue meaningful goals.[2]

O
Oxytocin
THE CONNECTION CHEMICAL

Oxytocin connects you with the people you love and builds your self-belief.[3]

S
Serotonin
THE MOOD & ENERGY CHEMICAL

Serotonin creates huge, positive shifts in your mood and energy, and drives you to pursue healthy behaviours.[4]

E
Endorphins
THE DE-STRESSING CHEMICAL

Created by physical movement, endorphins de-stress and calm your brain.[5]

First, let's consider how we humans have spent 99.9 per cent of our time here on earth.

We started off deeply immersed within the natural world, surviving and thriving as tribal communities. For context, it is estimated that, for the majority of human history, we spent 85 per cent of our time outside. Now in the modern world we are spending just 7 per cent of our time outside.[6] It is fascinating to imagine how our ancestors' brain chemistry would have been forming alongside this lifestyle. Their **dopamine** levels, the motivational chemical that is built through hard work and effort, would have been surging, given their constant challenging pursuit of survival. Their **oxytocin** levels, the connection chemical, would have been rising every day, given how essential it was for them to remain connected as a group in order to survive. Their **serotonin** levels, the mood and energy chemical, would have been booming given their days were spent outside, in nature, in the sunlight, eating unprocessed foods. And their **endorphin** levels, the de-stressing chemical, which is created through physical movement, would have been soaring through building, hunting, and surviving as a group.

For most of history, we spent	In the modern world we are spending just
85%	**7%**
of our time outside	of our time outside

Now let's imagine we put hunter-gatherers into today's world.

Suddenly they have access to sugary processed foods and their **dopamine** levels begin to drop. They are constantly distracted by phones and social media and their **oxytocin** levels fall. They begin spending all their time inside and stay awake late at night, and their **serotonin** levels decline. They become sedentary, sitting behind desks all day, and their **endorphin** levels reduce.

This is where we are at as a society right now. Many of us have lifestyles that prevent us from producing enough of these essential chemicals. Once you understand this, however, there is a simple answer. And you are holding it in your hands right now.

THE ANSWER: *The book you're holding in your hands right now.*

Your DOSE journey begins here

In each part of *The DOSE Effect*, you'll explore one of the four key brain chemicals. For each chemical, you'll discover five DOSE Actions – science-backed activities designed to optimise that particular chemical. I recommend trying one DOSE Action each week. Soon, they will start to become an essential part of your day-to-day life. This process is going to be achievable and incredibly impactful. Our world is magnificent, and the progress we are making as a society is phenomenal. Hunter-gatherers could only have dreamed of the world we live in today. We must now dream back to their world, and find a balance between who we instinctively are as humans, and who we are becoming in our ever-advancing world.

In order to get the most out of this book and truly transform your life, there is one pivotal impulse you must immediately begin to connect with. I am talking about the feelings and the messages you hear from your brain and body each day. Now that may sound unusual at first! But I am sure you feel emotions in your gut at times and in your brain at other moments. Feelings of dissatisfaction, of loneliness, of low confidence, of tiredness, of worry, and of stress. These feelings arise within you for a reason. They are here to guide you to make changes. Changes to the way you are living your life.

You must understand that your brain is a survival mechanism. Our brains have managed to help us survive as a species for over 300,000 years, and that wasn't easy. Take a moment now to imagine attempting to survive outside through the cold winters without modern comforts. It's phenomenal that we are still here. The reason we are is because of powerful instincts that have guided us towards survival. They made us want to hunt, to care for our children, to build shelter, and to continue to innovate in every aspect of our lives in order to thrive. When we engaged in these behaviours, we experienced rewarding feelings within our brain and body that made us want to do them more. These rewarding feelings were our brain chemicals activating.

Now our lives are very different. But our brain chemistry operates in the same way. It is still trying to guide us, just as it was rewarding us for the behaviours that created the optimum conditions for survival. It is now doing the same with the behaviours that are causing our decline. For example, when you procrastinate and scroll on your phone for hours and then feel depleted and demotivated afterwards, it isn't a coincidence. This is your brain knowing that scrolling is not the path to the optimal future for you. So your brain will make you feel awful in order to guide you to adjust your behaviour. The same thing happens when you eat tons of sugar, spend too much time inside, sit down all day, drink too much alcohol, or watch too much porn. All of these actions, which have become a normal part of modern living, reduce our potential as a species. Therefore, our brain chemicals will continue to send us negative messages until we listen, until we make a change.

Very simply, you need to start listening. Listening to how your daily behaviours are making your brain and body feel. Becoming more aware, for example, of how sugar and social media impact your mind. And then, as you begin engaging with the DOSE Actions and you start to rebalance your brain, I need you to listen to how this feels, too. Notice the rise in motivation, the feelings of confidence, the boost in your mood, the relaxation in your mind. If you observe the changes, your desire to pursue a healthier life becomes easier. The motivation starts coming from within and these habits won't feel like a chore. They'll feel like a gift instead.

I have built and designed this book in a very specific way, based on how our brains operate. Throughout *The DOSE Effect*, you will come to understand the function of each specific brain chemical; you will discover what it feels like if that chemical is low or high, and identify the key causes that create these experiences. Then, and most importantly, you will begin completing challenges. This book is about true behavioural change. The challenges are specific and achievable. Your lifestyle will gradually shift, and this shift will cause a transformation in how your brain operates each day.

The DOSE Effect

I promise, you can achieve the goals that are most meaningful for you if you follow this formula. You will feel confident in who you are and connected to those you love; wake up feeling energized and ready for your day and you will truly feel peaceful and relaxed. I am so happy you are here. A new life is on the horizon.

Introduction to
TJ Power

Hi, my name is TJ Power and I will be guiding you through your DOSE journey. I am super-excited that you are here. Given the connection we are soon going to build throughout this book, I feel it would be valuable for you to know a little about my story so far and what has brought me to this moment.

I am someone who struggled greatly at school. I found it very hard to connect with subjects that I didn't find interesting. At the age of sixteen, I went to a college in the UK and discovered psychology for the first time. I immediately fell in love with it and wanted to learn more. During the following five years, I navigated some challenging times as a young man. I lost five people, both young and old, who I was incredibly close to. By the age of twenty-one I had carried four coffins on my shoulder. This time in my life matured my mind and I often felt extremely low. During this period, I turned towards the party lifestyle, which is so common to do. Unfortunately, I began to engage very regularly with what we will come on to discover as addictive dopamine behaviours. I did this both in the pursuit of fun but also in the pursuit of avoiding my emotions.

Meanwhile, I continued to love my study of psychology, and the summer before I started my Master's, I decided I wanted to get my mind and body healthy for the first time. In order to do this, I chose to go and live with my grandpa in the countryside. I took the time to focus on listening to my mind and recognizing my instinctive feelings. This began a new path to a happier and more focused life. I was simply trying to sort out my depressed and addicted mind.

During my Master's, I focused on my journey to becoming a neuroscientist through the study of health science, psychology, and neuroscience. I was extremely fortunate, at the age of twenty-one, to be offered the opportunity to create my own lecture series at Exeter University, where I was on stage talking about the research behind the topics I loved – psychology, neuroscience, flow state, and much more. My journey towards becoming a respected neuroscientist quickly gained momentum. I spoke at Oxford University a couple of months later, and many more universities after that. During this time, the idea behind *The DOSE Effect* was coming to me: a simple, science-backed formula that was easy to understand and actionable by anyone. A formula that focuses on the positive. It provides people with their north star – a goal and direction towards where they want to be. I quickly discovered that it was so much better if people were pursuing the idea of feeling good, rather than running away from feeling bad.

Following my Master's, and lecturing, I co-founded the company Neurify alongside becoming the lead neuroscientist at our research centre, The DOSE Lab. Neurify is a major international mental health and performance training company. Throughout the early years of Neurify, I delivered DOSE Live, a lot. I personally trained over 50,000 people within two years and couldn't believe the results. We began testing people on key metrics before and after their training, and the results were staggering.

Some of our findings included:

48% — IMPROVEMENT IN CONCENTRATION AND DEEP FOCUS

49% — IMPROVEMENT IN MOTIVATION THROUGHOUT THE DAY

50% — IMPROVEMENT IN HEALTHY PHONE USE

59% — IMPROVEMENT IN DAILY ENERGY LEVELS

30% — IMPROVEMENT IN SELF-BELIEF AND POSITIVE SELF-TALK

54% — IMPROVEMENT IN SLEEP QUALITY

32% — IMPROVEMENT IN ABILITY TO NAVIGATE HIGH STRESS

29% — IMPROVEMENT IN ABILITY TO NAVIGATE ANXIETY

40% — IMPROVEMENT IN MOTIVATION TO EXERCISE

41% — IMPROVEMENT IN HEALTHY NUTRITION

*Data recorded use a pre and post 1–7 Likert scale measure, before and after

I couldn't believe it. We had a method that everyone appeared to understand. One that resonated deeply with people and one that was truly working. With this, I got focused. Even more focused than I had been before. We began building an incredible team, an amazing technology platform, physical products, and this has brought us to today, and the greatest honour in my life so far: the opportunity to write this book. After over a decade of deep research, alongside teaching thousands of people, I have put everything into *The DOSE Effect*.

It will provide you with the clearest solution imaginable to truly thrive in our modern world. I hope you find it fun and interesting and, most importantly, I hope it has a transformational impact on your life.

Thank you for being here.
Let's begin.

TJ

PART 1

Develop the Ability to Achieve Your Goals

Understanding DOPAMINE

Dopamine is the brain chemical that has become incredibly well known in our modern world. You may be familiar with the idea that you receive small dopamine 'hits' when scrolling social media on your phone, drinking alcoholic drinks, or eating sugary foods.[1]

While this is true, dopamine is responsible for far more than this. Learning how to increase the amount of dopamine you produce naturally each day will have a monumental impact on your motivation levels and, with that, your capacity to pursue what you are seeking in your life.[2]

You need to learn how to naturally increase the amount of dopamine you produce each day.

The Dopamine Principles

PRINCIPLE 1:
MAKES HARD WORK FEEL GOOD

First we must understand the primary function of dopamine. For our former hunter-gatherer selves, dopamine was responsible for creating the drive within us to complete the hard tasks that would keep us alive.[3] Let's take the vital daily activity of hunting for food as an example. This was an activity that required a huge amount of motivation and deep focus to achieve. Dopamine would rise within our brains and create a strong desire to find food.[4]

Then, during the pursuit of hunting for it, dopamine would continue to rise as we got closer to the goal. Upon successfully hunting down an animal, dopamine would rise once again, creating a huge experience of reward and joy in our brains. This rewarding feeling would then strengthen our desire to complete this challenging activity on a regular basis, therefore maximizing our likelihood of survival.[5] The key to understanding dopamine is that, to get it, you must focus on completing tasks that at first require effort, then gradually create a feeling of progress, thus making you really feel good after. A simple example of this could be tidying your home[6] – an activity that is easy to put off and avoid with procrastination. However, once you eventually do it, it leads to a feeling of satisfaction and accomplishment. This feeling occurs as a result of the rise of dopamine in your brain.

PRINCIPLE 2:
CONTROLS THE PLEASURE–PAIN BALANCE

Now, let's understand how our modern world is messing up this vital brain chemical. An incredibly simple and effective way to understand this is through a brilliant concept popularized by Dr Anna Lembke in *Dopamine Nation*, called the 'Pleasure–Pain Balance'.[7] Recent neurological research has shown that the parts of your brain that experience pleasure and pain are 'co-located'. This means they are located right next to each other in your brain, in an area called the hypothalamus.

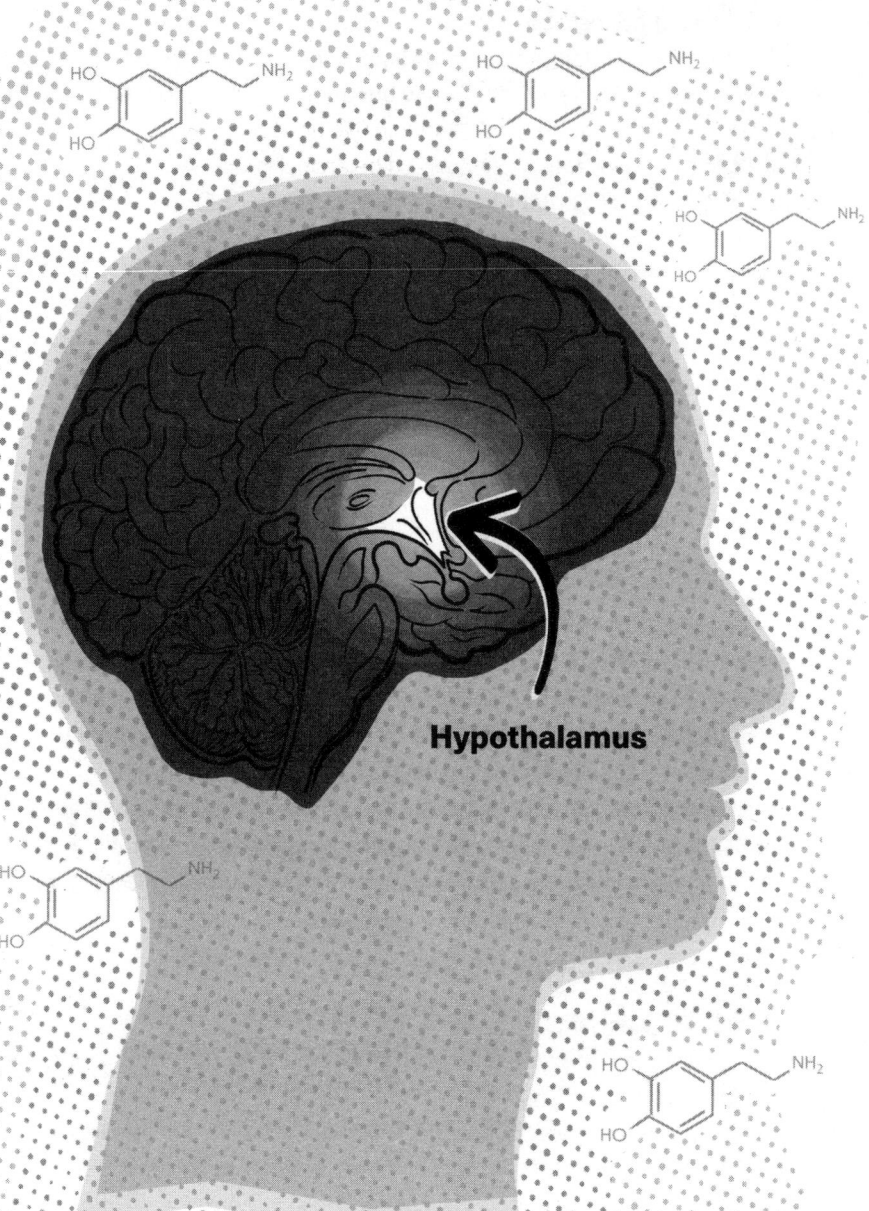

This is particularly interesting as, given their co-location, they operate as a see-saw. This means that when you do hard, 'painful' activities that result in either mental or physical strain, such as pushing yourself in the gym or concentrating for a prolonged period of time when working, the see-saw will be weighted on the pain side. Let's return to our ancestors to understand this better. Take a moment to imagine spending five hours outside, in the cold, searching for food and shelter. This would be an incredibly challenging activity. Given this difficulty, it was vital that our brains developed a survival mechanism, which ensured completing hard activities actually made us feel good. With the see-saw weighted on the 'pain' side, the 'pleasure' side of the see-saw would rise, creating a rewarding, positive feeling in our ancestors' minds. This would then lead to a reinforcement of these 'painful' activities, which were essential to our survival.

Originally, the only way to increase our dopamine levels was through these hard, 'painful' activities such as hunting, foraging for food, building shelter, creating fire, and finding somewhere to live. However, over time, we quickly found ways to stimulate our dopamine system without any effort, via cigarettes, alcohol, drugs, pornography, junk food and, now, social media.[8]

Let's return to the pleasure–pain balance see-saw. In a similar way that 'painful' activities resulted in the brain experiencing 'pleasure', these activities will tip the see-saw in the opposite direction, resulting in the brain experiencing 'pain'. This is simply evolution at its finest, helping us to survive. It's genius that we have a mechanism within our brains that rewards us when we engage in the key behaviours that increase our likelihood of survival, and makes us feel bad when we engage with those that reduce our likelihood of survival.

During these highly pleasurable dopaminergic activities, your brain also produces an incredibly intelligent additional neurochemical called dynorphin. In order to further dissuade you from engaging too much in them, dynorphin is released, creating discomfort in your brain.[9] This discomfort may be experienced as depressive feelings and severe low mood – the kind of feeling that occurs the day after drinking too much alcohol, eating too much sugary food, or scrolling through social media videos for too long.

Throughout *The DOSE Effect*, you will discover how to become a modern-day hunter-gatherer, progressively incorporating key activities that will tilt this pleasure–pain balance see-saw towards the side of natural pain, which is more sustainable and will lead to a happier and more motivated you. Our goal is no longer the pursuit of hunting animals or building shelter; now, throughout this journey, your pursuit will become your dreams, your passions, your relationships, and the health of your body and mind.

Do you have low dopamine levels?

Now you have a clear understanding of dopamine, I need you to first figure out what the primary behaviours are that you engage with regularly, which may cause low dopamine. We will refer to these behaviours as quick dopamine. In a low dopamine state, you will initially feel demotivated and you will procrastinate.[10] If you repeatedly exhaust your dopamine system with these behaviours for prolonged periods of time, you will experience low mood and depressive symptoms.[11] The six primary behaviours that will cause your dopamine levels to reduce are:

The SIX Main Causes of Low Dopamine:[12]

1. SUGARY FOODS
2. ALCOHOL AND DRUGS (INCLUDING VAPES)
3. PORNOGRAPHY
4. SOCIAL MEDIA
5. GAMBLING
6. ONLINE SHOPPING

If you read that list and think, wow, I engage with many of those behaviours on a regular basis, it is important to know that this is very common in our world. From a young age, I found all of these behaviours incredibly enticing and addictive. I began engaging with them all as a teenager and have been on an interesting journey of discovery throughout the last ten years to learn how to manage them. If I think back to a typical day at university, for example, I would wake up and immediately start scrolling on my phone, before finally convincing myself to get out of bed. On my way to classes with music or a podcast playing in my headphones, I would continue to frantically check my phone. I would then grab something fast and unhealthy to eat. I'd sit down at my laptop and attempt to do a little work, not much before I was distracted by YouTube videos or messaging my friends to help me to not feel bored and alone. I'd head to a lecture and find it impossible to concentrate. I'd head home so I could start drinking early with my friends and often end up partying till late in the night. I'd wake up feeling anxious, and then simply repeat this day again and again.

I empathize massively with the fact you're in a modern world with all these incredibly addictive temptations, no matter what stage of life you are at. That's why I feel I can guide you on this topic. Throughout this book, I will provide you with the most valuable insights and strategies I have learned, so that I can help you navigate our highly addictive dopaminergic world.

Before we go into the causes of low dopamine, I want to visually demonstrate how these activities negatively impact your dopamine levels. As you now know, dopamine is a chemical that is designed to be slowly 'earned' through effort.[13] Take the moment you are in right now, for example, reading this book. This is a perfect example of the opposite of these quick dopamine activities; we refer to reading as slow dopamine. When you open this book and begin reading, you don't immediately experience a surge of pleasure in your brain. However, after you have concentrated for a period of time, maybe five to ten minutes, you will notice a feeling of satisfaction and accomplishment arise. This is because you are 'earning' pleasure. Take a look at the graph on the next page to see how this looks.

In this graph you can see your dopamine levels on the left (the Y axis). You have your baseline level, the amount of dopamine your brain is naturally producing at any moment. You then have low dopamine below this and high dopamine above it. Going left to right you have the X axis representing time; this is the amount of time it takes for your dopamine levels to increase.

In this example of reading, your dopamine levels slowly rise as you read, as a result of your brain engaging in effort. A positive feeling of reward and satisfaction is then created. After you stop reading, your dopamine levels slowly fall back to your baseline.

Effect of Slow Dopamine

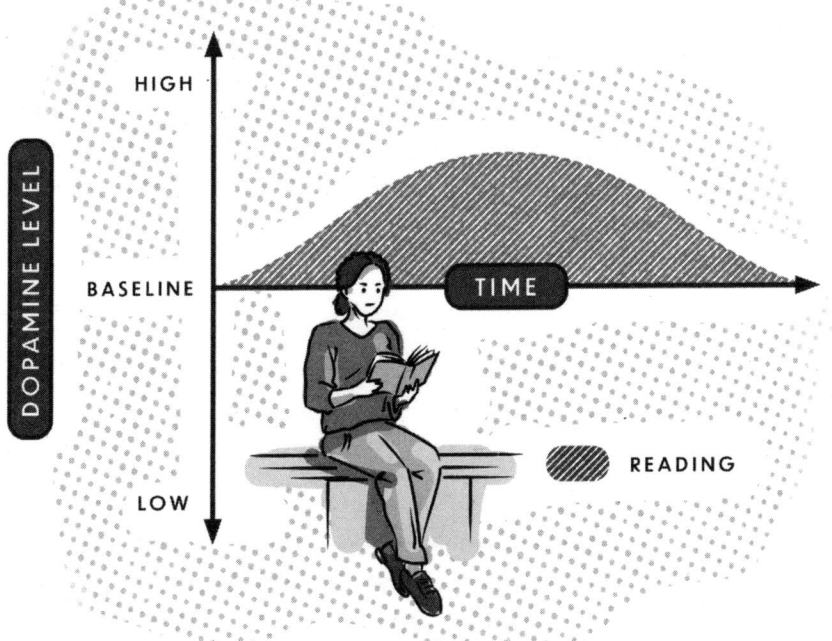

Now let's look at what is occurring in your brain when scrolling on social media, a quick dopamine activity. You open your social media apps and immediately you notice you feel really good, your dopamine levels increase incredibly fast and pleasure is experienced in your brain.[14] The challenge this creates is as with everything in our universe: as the laws of physics explain, 'what goes up, must come down'. Your brain and body are always seeking something called 'homeostasis', which simply means balance. With this in mind, when dopamine levels increase incredibly fast when scrolling social media, the brain then thinks, 'Wow, how are my dopamine levels so high?' In order to achieve homeostasis, or balance, the dopamine then has to quickly drop an equal amount below your baseline level in order to rebalance, making you feel even worse than before you started scrolling.[15]

Let's take a look at this second graph. You can see that as soon as social media was opened, dopamine rose within the brain. This is why scrolling is so tempting and so addictive. For our ancestors, it could take four hours of hunting to achieve this kind of dopamine increase. Now we can experience it instantaneously whenever we want it, rather than the four hours of earning the slow dopamine as they would have needed. In this graph you can see that once the phone is put down, the low dopamine symptoms of demotivation, procrastination, and low mood rise.

Again, remember, this is only your brain seeking to help you. It knows that lying in bed for two hours scrolling instead of going to sleep is not helping you thrive as a person. When you learn to listen to your brain's guidance instead of resisting it, a magical new experience of life awaits.

Effect of Quick Dopamine

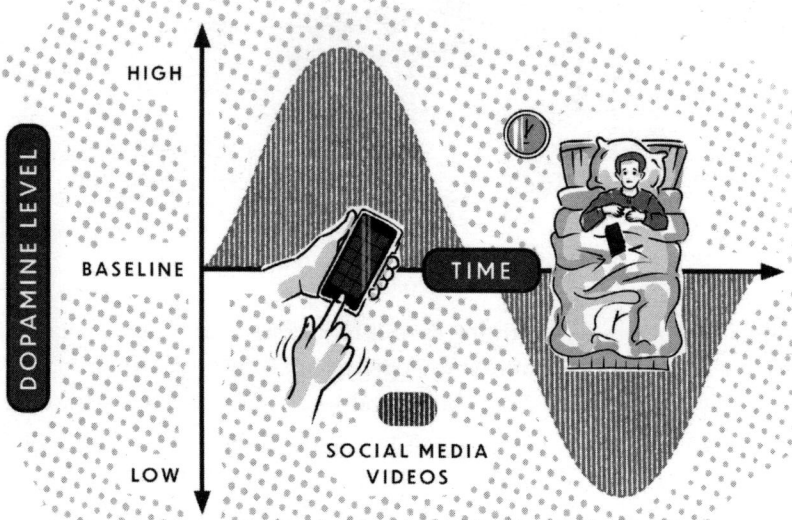

When you repeatedly spike and crash your dopamine levels, it exhausts your dopamine system. If you got in your car every morning and revved the engine for five minutes without putting it into gear and driving it, you know this would burn the engine out. In a similar way, many of us are now burning out our dopamine systems with pleasurable behaviours. This is a primary cause of the demotivated, depressive symptoms you may experience in your life.[16]

Any of the aforementioned behaviours (sugary food, alcohol, drugs, pornography, gambling, and social media) result in significant spikes and crashes in your dopamine system. These quick spikes in dopamine are what make these behaviours incredibly addictive.

Andrew Huberman, the extremely popular neuroscientist, defines addiction as 'the progressive narrowing of the things that bring you pleasure'.[17]

I want you to re-read that sentence and think about whether you are experiencing this. I certainly was for much of my life. 'The progressive narrowing of the things that bring you pleasure.' You may be finding that a lot of the pleasurable moments in your life happen largely while scrolling through your social media feeds, eating sugary foods, or drinking alcohol. Again, this is normal and nothing to judge yourself for. The challenge this creates is constantly having your dopamine hacked, as these behaviours deplete the one chemical in your brain that is going to provide you with the drive to create the life you want to have. When we look at the alternative side of Huberman's phrase, 'a good life is the progressive expansion of the things that bring you pleasure', we can begin to explore a new way of living.[18]

We want to find joy beyond these quick dopamine behaviours and instead discover more natural forms of happiness.

Natural Forms of Happiness are:

1. MAKING DEEP SOCIAL CONNECTIONS
2. MOVING YOUR BODY
3. EATING NUTRITIOUS FOODS
4. READING BOOKS
5. SLEEPING DEEPLY
6. TAKING CARE OF YOUR HOME
7. WORKING TOWARDS YOUR GOALS

Over the course of this book, I am going to help you to counter the negative effects of our modern addictive world and show you how you can discover true pleasure in the natural joys the world has to offer.

How high dopamine levels will feel

As you engage more often in effortful activities, your capacity to experience high dopamine will increase.

In a high dopamine state, you will feel two key symptoms. You will feel productive – you will find your motivation levels are higher, and completing the activities in your life that you know are good for you will feel far more achievable.[19] You will also experience the emotion of excitement more.[20] You know that feeling, when you are genuinely excited about your life? This is high dopamine. A really good way to understand this is by looking at how we feel in our average week. Typically you may find that on a Monday morning you will be quite the opposite of excited; instead you feel a little sluggish and unmotivated. In comparison, on a Thursday and Friday you may feel more excited by your life. As a society we think this is occurring simply because the end of the week is coming. However, it is more complex than that. Typically on a Friday, Saturday, and Sunday, we will engage far more with quick dopamine. We will drink more alcohol, eat more sugar, and scroll our phones for longer. This will then lead to low dopamine on a Monday morning, hence the lack of excitement, motivation, and ability to concentrate. Then, throughout Monday, Tuesday, Wednesday, and Thursday, we engage in more effortful, slow dopamine behaviours. This may be working hard at your job, maintaining an organized home or going to the gym more often. As we engage in this effort throughout the early part of the week, our dopamine builds, leading to the feeling of excitement and motivation.

Feeling excited about your future is essential for your mental health. We will discover precisely how you will achieve this as we journey through *The DOSE Effect*.

When looking at your capacity to both resist engaging with the addictive behaviours you find pleasurable, and ensure you incorporate the effortful daily activities that you know will benefit your future, we must consider how strong your willpower is. The science of willpower is fascinating and something that we all desperately need to strengthen in order to thrive in our modern world. Willpower can be defined as your ability to resist short-term temptations in order to achieve your long-term goals. To give you a simple example of this, if you decided your long-term goal was to eat healthier food and improve your physical health, then resisting chocolate and biscuits and selecting natural foods would serve this mission of yours.

Now the science: there is a specific area of your brain called the anterior mid-cingulate cortex, or AMCC.[21] During any moment in which you either resist an addictive behaviour or intentionally engage with a healthy behaviour, this component of your brain will light up. What is fascinating here is the more often it lights up, the stronger it will become, in a similar way that going to the gym and doing bicep curls will strengthen your arms. With each rep, your arms get stronger. With each activation of your AMCC, your willpower will get stronger.[22] What's interesting is that as this component of your brain gets stronger, your ability to maintain discipline gets easier. If, for example, you finish work and think, 'I should go to the gym', and your brain counters with, 'Ahhh, but I can't be bothered', but you still you force yourself to do it, your AMCC will light up and strengthen, which in turn makes the next time this negotiation occurs in your mind a little bit easier to win. You know when you meet one of those really disciplined people, who eats well, exercises lots, works hard, and you think, 'How on earth are you doing this?' That person has activated their AMCC.

QUICK ACTIVITY

I want you to select an addictive dopamine behaviour in your life that you will use to begin strengthening your AMCC and willpower.

First select your behaviour:

1. **SUGARY FOODS**
2. **ALCOHOL AND DRUGS (INCLUDING VAPES)**
3. **PORNOGRAPHY**
4. **SOCIAL MEDIA**
5. **GAMBLING**
6. **ONLINE SHOPPING**

Throughout the rest of this week, in each moment that you get an urge for this behaviour, I want you to think about this section of the book. Mine at the moment is social media. As I write this book, at this very moment, I have a strong urge to stop writing, pick up my phone, scroll, and get some quick dopamine. I am reminding myself that each time I resist it, my AMCC lights up, my willpower gets stronger, and my ability to stay in deeper states of focus improves. Yours may be sugary food, porn, or alcohol. Remind yourself that each time you successfully resist it, that's a rep in the gym for your brain. The less you engage with these quick dopamine behaviours, the more motivated and positive your brain will become.

Congratulations on reading the introduction to dopamine. It is not easy to prioritize reading over other activities in your life. Celebrate this as an achievement. You are now taking significant steps towards a more positive future. Our modern world is full of distraction and pleasure. Those who develop the discipline to manage it will thrive. You are on the journey to becoming one of those people.

On the next page, you will find a summary of the key functions, principles, feelings, and behaviours that are associated with optimizing your dopamine.

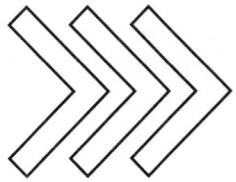

Dopamine Summary

Function →	• Motivation • Concentration
Principles →	• Makes hard work feel good • Controls the pleasure–pain balance
Low Dopamine Symptoms →	• Demotivated • Distracted • Depressed
Low Dopamine Causes →	• Sugary foods • Alcohol and drugs • Pornography • Gambling • Online shopping • Social media
High Dopamine symptoms →	• Motivated • Determined • Excited
Dopamine actions →	• Flow state • Discipline • Phone fasting • Cold water • My pursuit

1

Develop Your Powers of Concentration

DOPAMINE
FLOW STATE
DOPAMINE
FLOW STATE
DOPAMINE
FLOW STATE
DOPAMINE
FLOW STATE
DOPAMINE
FLOW STATE
DOPAMINE

First, let's rate your attention span.

Rate yourself on a scale from 1 to 10 for how good you are at concentrating. Be honest with yourself.

1 → 10

1 = terrible
10 = amazing

Understanding FLOW STATE

Wow, hasn't it become hard to concentrate these days? Whether you're in a conversation with someone, working on your computer, or even watching TV, it has become incredibly challenging to get our brains to focus for prolonged periods of time on just one activity.

This is hardly surprising, as there is an opportunity to distract yourself in every moment. As a society right now, we are almost just accepting this and have given up trying. You hear people saying, 'I literally just can't focus any more, it's impossible.' However, this doesn't have to be your reality: your ability to concentrate is something you can train your brain to do, just like any other skill. It is absolutely essential to your future that we redevelop your ability to achieve deep states of focus.

There is a very specific reason that your first DOSE Action is **Flow State**. This is because the daily act of reading this book is one method in which we will train you to achieve it. Flow State is a concept I began deeply researching and writing about during my Master's and fell in love with. It is a concept initially developed and popularized by the Hungarian-American psychologist, Mihaly Csikszentmihalyi.[23] It refers to the state of deep focus that occurs when someone becomes incredibly immersed in an activity. During Flow State, individuals lose their perception of time, disconnect from distraction, and reach new levels of performance and productivity.[24] If we consider our ancestors again, the levels of Flow State, or deep focus, they would have experienced during hunting, foraging, building, or fighting would have been unbelievable. Flow State is something that we would have experienced far more regularly just a few decades ago, prior to the level of technological distraction we experience today. Nowadays, any time that we try to deeply focus our minds when working or even concentrating on a TV show at home, we distract ourselves long before these states of deep focus can begin.

Flow State has an incredibly intricate relationship with your dopamine system.[25] When we are in these deep states of focus, our brain is engaging a great deal of effort, and as a result your dopamine levels increase significantly.[26] Let's take the moment you are in right now as an example, reading this book. Your brain is engaging in effort in order to absorb these words. Far more effort than if you were scrolling through your social media feed. You will notice that the more you deeply engage with reading this book, the faster and more effortlessly you will begin to achieve this. This is your brain beginning to enter Flow State, where it is progressively letting go of the distraction around you and the distracting thoughts in your mind, and is becoming more and more focused on the words on this page. What is interesting with focusing on a task, whether it's this book, or a project you are working on, as you get past the initial desire for distraction and remain focused, the more you begin to 'gain momentum'. This means your brain begins to operate and process information faster and faster.[27] You may have experienced this in your life before, potentially when you had to complete a project, if there was a deadline and you reached a point where you just had to get it done. You cut the distraction, zoned in, and achieved a heightened level of productivity. You will notice when you work in this way, or read in this way, you experience a far more rewarding, satisfying feeling in your brain afterwards. This is because you have begun to enter Flow State and dopamine levels have slowly risen in your brain.[28]

I want you to take a moment now to consider where you experience Flow State in your life at the moment. It is a state that you may experience when working on your computer, but it can be achieved in a variety of other activities too.

Flow State is an experience the human mind deeply loves and craves. It is incredibly important that you strive to identify where it occurs in your life and how you can make more time to experience it. This will support you as you optimize your dopamine levels and build a progressively rising level of motivation in your life. The additional beautiful side of Flow State is the relief it provides us with from our more anxious or worrisome thoughts. As we will discover in Chapter 14, Underthinking, overthinking is a challenge many of us experience. Flow State activities immerse our minds so deeply in the present, in the task in front of us, that overthinking, worrying about our past or fearing for our future diminish and a peaceful productivity arises in its place.

You may enter a FLOW STATE when:

1. RUNNING OR GOING TO THE GYM
2. PLAYING A MUSICAL INSTRUMENT
3. PAINTING OR DRAWING
4. WRITING OR JOURNALLING
5. CODING OR PROBLEM SOLVING
6. GARDENING
7. CLEANING
8. READING

Strategy

As fantastic as this all sounds, you may now be thinking, TJ, I literally cannot focus, I cannot stop picking up my phone; my brain gets so bored and craves distraction. Please bear in mind, I am with you on this; over the last three years, I have had to follow all the techniques I share in this book to develop my ability to concentrate. Just a few years ago, I could never have concentrated long enough to even write this book in the first place. Flow state is not a skill I am blessed with but it's one I need for my work, and I know my brain and mental health benefit massively from it. I have deeply researched the neuroscience of focus alongside exploring many strategies to enhance my ability to concentrate – many of which I am using at this very moment as I write this book . . .

Despite the urge to pick up my phone and check my social media, I am now going to take you through a simple 4-step guide to training your concentration in order to enter Flow State. We are going to specifically focus on your working life, as this is where Flow State will have significant benefits for you. However, given what we have discussed about the wide range of activities where Flow State can occur, I will guide you to consider elements of this strategy in any area of your life.

STEP 1:
SELECT THE TASK

First, you must carefully select precisely which specific task will be completed, ensuring it is one that is achievable in the time period you have available. Ensuring the task is challenging but achievable creates the feeling throughout the concentration session that you are getting closer towards your desired goal. This feeling that you are approaching the completion of a goal will further increase your dopamine levels in order to keep you motivated. We will discover the importance of this more deeply in Chapter 5, My Pursuit. In order to enter Flow State, we must have a very specific goal in mind that we would like to achieve. Remember, dopamine is built through effort.[29] The harder the task, the more effort required and therefore the more dopamine that is created. If you can attempt to tackle a challenging task in the morning, this will create significant increases in your dopamine levels throughout the rest of the day.

One challenge many of us have is a huge to-do list. These huge lists can overwhelm us and lead to 'procrastiscrolling', the state where we just ignore our tasks altogether and sit on our phones. We must learn to avoid this. When you are working you may do the opposite of selecting one task; you may try to gradually work through all of your tasks at once, jumping around at any moment if the present task is boring or difficult. Task switching in this way has been shown to reduce your overall productivity by 40 per cent.[30] That means you are almost making every single task take double the amount of time it requires. Once your task is selected, move to step two.

STEP 2:
TELL SOMEONE

Accountability is a powerful way to commit yourself to achieving your desires. As you journey through this book, I will be guiding you to share with people that are close to you the DOSE challenges that you will complete, in order to increase the likelihood of achieving your goal. So, once you have selected the task you are going to complete, you need to tell someone. This means, for example, messaging a co-worker or a friend and saying, 'I am going to focus on completing that slide presentation now.' This will commit you to the task and ensure you start right away. Once you have shared your selected task with someone, move on to step 3.

STEP 3:
ELIMINATE DISTRACTION

One of the hardest components of concentration is getting bored or finding a task too difficult. It is in these moments that we distract ourselves, by clicking on something else on our computer... our emails, company messaging apps, social media, and so on. Once you have selected your task and shared with a colleague or friend that you are about to tackle it, you must close ALL distracting apps on your computer and not open them again until your chosen task is complete. Once all distractions are removed, move on to step 4.

STEP 4:
THE STOPWATCH CHALLENGE

Now, the final part and the most impactful. We need to train your brain to concentrate for prolonged periods of time in order to enable you to enter Flow State and achieve true productivity. It is really important to understand that the beginning of a task is the hardest, and it's the hardest by a long way. As we discovered earlier, once you are deeply in a state of focus, your brain gains momentum. In order to achieve this momentum, we must ensure you can surpass fifteen minutes of focus. These initial fifteen minutes are the hardest, as this is when your dopamine levels are low and only just beginning to rise. This is when you are most vulnerable to distracting yourself. Pushing through this discomfort at the start is the key to transforming your focus and productivity.

In order to stick with the activity, as you start your chosen task, open your phone and click on the clock app, and then the stopwatch. Important tip: ensure airplane mode is on to avoid any notifications coming through! Once you start the task, click start on the stopwatch, place your phone in another room, face up, and it will begin counting 00:00, 00:01, 00:02, 00:03.

Return to your working area and zone in. What you discover is very quickly you find yourself wanting to distract yourself. If you go over to your phone and see it says 01:07, you will think wow, I can only concentrate for one minute and seven seconds. You then say to yourself, right, I must be able to go longer than this. You get back into the task and push through. Then a few moments later you may return to the phone and see it says 08:37. Okay, brilliant, you've now focused for eight minutes. You can now begin gamifying this process with yourself, seeing every day whether you can progressively increase this number and retrain your ability to focus. First your goal is fifteen minutes, then you should increase it to thirty minutes, then forty-five minutes. What's amazing is that once you reach fifteen minutes, you then know you are beginning to enter Flow State, and avoiding the distractions becomes easier, because you think, 'I've just put in all this effort to get my dopamine up: don't crash it now!'

Flow State Summary

STEP 1 Select the task

STEP 2 Tell someone

STEP 3 Eliminate distraction

STEP 4 The stopwatch challenge

Flow State activities immerse our minds so deeply in the present, in the task in front of us

Dopamine & ADHD

When considering our ability to enter these deep states of focus, it is essential we also consider the ever-growing modern challenge of attention deficit hyperactivity disorder (ADHD). ADHD is on the rise in our modern world. A component of this will, of course, be our advancing capability to diagnose it. However, we must also explore how quick dopamine is impacting its prevalence and associated challenges.[31]

An individual navigating ADHD will experience two things; one is inattentiveness, a reduced ability to remain focused on a task, and the other is impulsiveness, an increased desire for stimulation and pleasure.[32] Individuals with ADHD have a reduced baseline dopamine level.[33] If we think back to our dopamine graphs on pages 29 and 30, their baseline may be more like the one you see below.

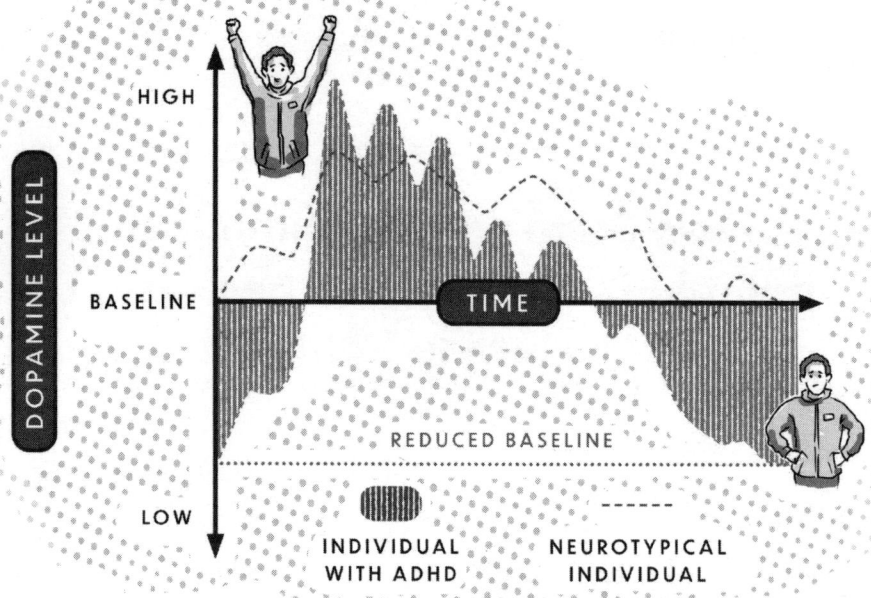

Dopamine and ADHD

This means that they can experience greater 'highs' from quick dopamine behaviours, making them more enticing. It also means they have a greater susceptibility to experience low dopamine and they struggle with procrastination and lack of focus.[34] For decades we have known that ADHD is intricately connected with dopamine, which is why the current medications for ADHD are drugs such as Adderall or Ritalin, which are designed to help elevate one's baseline dopamine levels.[35]

For many years there has, of course, been a powerful genetic component to ADHD, where individuals are predisposed to lower baseline dopamine levels. However, with the rise in accessibility of quick dopamine, we now have the capacity to create these constant surges and crashes in dopamine. This leads to many of us experiencing low dopamine symptoms and therefore exhibiting some of the ADHD symptoms, particularly lack of focus and impulsivity.

What is important here is that if you are someone navigating ADHD or you are displaying some of the symptoms, you must recognize how critical it is for you to begin managing your dopamine better. This means intentionally incorporating more activities into your daily routine that will help raise your baseline dopamine levels.

Important activities if you have ADHD Symptoms:

1. Having a structured morning routine that includes making your bed and stepping outside into natural light.

2. Exercising your body as often as possible, particularly before trying to focus on a challenging task.

3. Reducing consumption of sugary, quick dopamine foods, especially in the mornings.

4. Reducing length of time spent scrolling short social media videos.

FIND SOMETHING YOU LOVE

What is fascinating about individuals who are navigating ADHD is if they find an activity they love, they have the capacity not just to do it well but to really excel at it. A significant reason for this is because of their lower baseline dopamine level, meaning they will experience greater rises in dopamine during this activity. This will make the activity feel far more rewarding and enjoyable, and is why individuals with ADHD can occasionally experience hyperfocus on activities they love.

With this in mind, it is incredibly important to experiment with different activities and identify one you love. As someone who grew up with many ADHD symptoms, I struggled for years with my inclination towards all the addictive behaviours of the modern world. Once I came up with the concept of DOSE, and found public speaking, writing, and making videos, it provided my mind with something I could really focus on to boost my dopamine and therefore access Flow State. Take a moment to consider what is the activity you love most, which provides the greatest feeling of reward. The activity you select is essential to thriving with an ADHD brain. Some activity ideas are in the box below.

ARTISTIC	EDUCATIONAL	SPORT-BASED
Drawing	Studying	Football
Writing	Problem solving	Swimming
Painting	Puzzles	Working out
Music	Listening to podcasts	Dance
Video		Golf
Craft	Reading	Yoga
knitting	Learning a new skill	Cycling

As you journey through *The DOSE Effect*, these insights will become clearer and clearer. A more disciplined, focused, and fulfilled brain is coming to your life very soon.

Challenge

I would now like you to complete the '**Flow State Challenge**'. In order to complete this challenge, you need to achieve one session of deep-focused work or your chosen activity every morning and one every afternoon for the next seven days. This may include zoning in on a key project that is important to you, working on a creative solution to a work-based problem, evolving an art project you are working on, clearing out your garage, or simply getting deeply focused when reading *The DOSE Effect* each day.

With your Flow State Challenge now selected, please be sure to tell someone you are close to that you will be doing this. This could be a friend, partner, or family member. This increases your accountability to achieve it alongside providing the opportunity for conversation with them about this topic.

KEY TIP:

Think about the science of willpower from our dopamine introduction when you are intending to enter Flow State. With each successful moment you avoid distraction and remain focused, your AMCC lights up, and your capacity to enter Flow State increases.

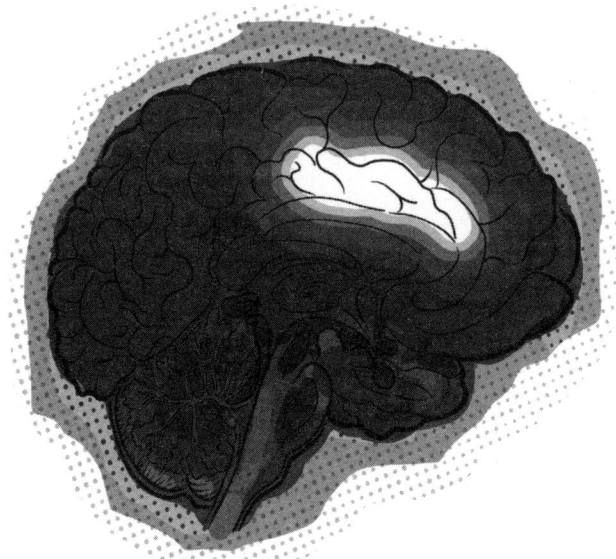

2

Build Discipline at Home

First, let's rate your discipline.

Here we are specifically referring to your discipline in your home environment. Rate yourself on a scale from 1 to 10 for how good you are at maintaining a tidy bedroom. Be honest with yourself.

1 ➜ 10

1 = terrible
10 = amazing

Understanding DISCIPLINE

Our external environment reflects our inner environment. Create a clear space to experience a clear mind.

Our next DOSE Action is one of the simplest and most important. In order to optimize your brain chemistry, you must develop your **Discipline**. Being disciplined is a fundamental characteristic of a happy, high-functioning person in the modern world. Being disciplined, however, in a world that is constantly fighting to pull you towards short-term pleasure and away from your goals, is incredibly challenging. During this next phase of reading, you will develop your self-control. The act of doing this will increase your dopamine levels,[36] alongside enhancing your ability to engage with all of the actions to come throughout your DOSE journey.

When looking at developing your discipline, a few interesting groups of people to consider are members of the armed forces, professional sportspeople, and monks and nuns, all of whom are unbelievably disciplined. When assessing how discipline is developed in these various domains of human excellence, it is initially instilled through how they engage with the environments they live in.

Our goal here is not just to strengthen your discipline but to help you develop a greater level of love and care' for the environment you live in, nurturing your home, your bedroom, your lounge, your work area. Making them calming and warm environments to rest, work, and socialize in, is important. As I grew up my mum did something incredibly clever that helped develop this ability to nurture my home environment. When I was a young boy, maybe five years old, my mum used to say to me, 'I love that you love keeping your room so organized.' Read that again. 'I love that you love.' From a young age my mind began to feel and believe deeply that I did love having an organized bedroom. As I got older this love became very real. Whether I am in a low headspace and need a boost of productivity to get me motivated, or I am in a more anxious, worried headspace and need to create more calm around me, I now genuinely love organizing my environment to help me achieve the headspace I am looking for.

DAY 1:
GET ORGANIZED: HOME

Have you ever noticed that when you organize and clean your home, despite it potentially being an annoying task, that once you are finished, you feel a sense of accomplishment and satisfaction? This is what dopamine is all about – engaging in effortful activities that add value to your life.[37]

Today your aim is simple: clean one room in your home. I advise choosing your bedroom. Your bedroom is an externalization of your mind. Clean it and observe what happens. The pursuit and accomplishment of cleaning it will increase your dopamine levels and clear your mind.[38] In order to achieve the greatest dopamine increase from this, I need you to really clean it. That could potentially mean opening your cupboard or drawers, throwing the contents onto your bed, and reorganizing it all. It might mean stripping your bed and cleaning your sheets or it might mean vacuuming the whole thing. Remember to take it one step at a time. For example, start with making your bed. Then clear away your clothes. Then organize the random stuff you have on your bedside table. Taking it one step at a time reduces any feelings of overwhelm if the task feels daunting. I would guide you to put some of your favourite music on (more insights on music coming in Part 4, Endorphins), and get this bedroom cleaning underway. Aim to create an environment in your room that makes you feel calm.

DAY 2:
GET ORGANIZED: WORK

After your bedroom is complete, we move towards your working environment. This may be an area of your bedroom or in another spot in your home. Wherever it is, it is incredibly important for your productivity that this area becomes, and remains, clean and organized. Take some time to reorganize it. This will further support our mission to help you achieve Flow State when you are working. Note: if you work in a profession that doesn't involve having a work environment at home, focus instead on organizing your kitchen.

Notice, as you make progress with this, the feeling of accomplishment and satisfaction that arises. Throughout the following few days, I want you to make it your primary focus to progressively organize your entire home one step at a time. I am aware that this is a significant task to undertake. Go at your own pace. If you live in a home with other people, be that your partner, friends, parents, or kids, it would be incredibly valuable to involve them in some way if you can. Just tell them you're reading *The DOSE Effect* and it's great for their dopamine too. I'm sure they could do with boosting theirs as well!

Collectively contributing to ensuring your home environment is organized is a simple act of contribution and kindness for others. Contribution is something we will be exploring in depth during Part 2, Oxytocin.

With your home now more organized, our aim is to maintain it that way. This begins with how you start your day.

 **The first thing I want you to do every day, immediately after waking up, is to make your bed.**

I want you to do this before you go on your phone, before you do anything. This simple act of discipline and accomplishment will set the tone for your day ahead. In the following chapter, Phone Fasting, we will be significantly altering your daily relationship with your phone, and later we will be carefully constructing the perfect morning routine for you.

With the task of making your bed now complete, I want you to then begin to look at all of your household chores in a new way. A way that understands the importance of task accomplishment and the impact this will have on your motivation levels. If, for example, you are in a low dopamine state, this may be a lazy, lethargic feeling, one where the idea of doing a work task is challenging. In a moment like this, organization of your environment is a perfect way to begin building your dopamine levels in preparation for a task that requires significant effort, for example prior to entering a Flow State activity. Each time you have to empty the dishwasher, wash your dishes, take the bins out, see these tasks as valuable to your mental health. I know this can seem unusual, but if you spend a week implementing Discipline in this aspect of your life, you will begin to see it translate to other areas.

Pay attention to whether you experience Flow State when you are cleaning. You may notice at the beginning it feels annoying and requires tons of effort. Gradually you begin to gain momentum. This is the start of Flow State. Dopamine is rising in your brain, making the task feel easier and more satisfying.

Strategy

The FOUR KEY AREAS to focus on maintaining discipline are:

1. MAKE YOUR BED EVERY MORNING
2. MAINTAIN AN ORGANIZED BEDROOM
3. MAINTAIN AN ORGANIZED WORK SPACE
4. REGULARLY WASH DISHES

Challenge

I would now like you to complete the '**Discipline Challenge**'. In order to complete this challenge, you need to deep clean your bedroom, work space, and kitchen during the next three days.

As you complete this challenge, be sure to tell someone you are close to that you will be doing this. A friend, partner, or a family member. This increases your accountability to achieve it alongside providing the opportunity for conversation with them about the value of engaging with this activity more often.

Break Your Phone Addiction

(3)

First, let's rate how healthy your relationship is with your phone.

Here we are specifically referring to how much time you spend looking at your phone every day. Rate yourself on a scale from 1 to 10 for how addicted you are to your phone. Be honest with yourself.

1 → 10

1 = terrible, the phone is always in my hand

10 = amazing, I check my phone occasionally

Understanding PHONE FASTING

Congratulations on reaching Action 3 in *The DOSE Effect*. Do not underestimate the power of what you are already achieving on this journey. You are now well on your way to optimizing your brain chemistry! The time has now come for us to discuss your relationship with your phone.

 **The phone is our specific focus here as it is the most accessible method of getting online. However, if a tablet is your primary method to engage with online content then this guidance applies to that device, too.**

This chapter will be one of the most pivotal aspects of DOSE, one that will have a huge impact on how you feel. First, we must recognize that your addiction to your phone is unbelievably common; it is not unusual that you love picking it up all the time, that you love scrolling, that you feel drawn towards social media so regularly. With your understanding of dopamine progressively increasing as you work your way through this book, we now know that the phone, particularly social media, provides incredibly fast dopamine hits.[39] These dopamine hits are unbelievably desirable and we have now invented a way in which we can quickly satisfy the deep dopaminergic urge within us by simply picking up our phone and opening social media. We also know, however, that these quick dopamine hits, which aren't earned through effort, will spike and crash our dopamine system, leaving us feeling demotivated, lethargic, and low.[40]

I have spent many years researching how we can develop a healthier relationship with our phones. Like all the guidance I provide in *The DOSE Effect*, this is something that resonates with me massively. I find my phone unbelievably addictive. I love my phone, and since I was a very young kid I have absolutely loved technology. I have loved gaming, iPads, computers, every item of tech I've engaged with. I remember getting my first iPhone when

I was in my early teens, the iPhone 3GS. It was awesome. At the same time, I got my first social media accounts and discovered the joy I could find there too, from connecting with people to sharing about my life. Progressively my addiction to these platforms began to increase. All of this was at a manageable level until TikTok was invented and boomed during COVID. The social media platform popularized the concept of short-form videos that you can quickly scroll through. If you take a second to think about your relationship with your phone, you may also notice that since the various COVID lockdowns your time spent on it has significantly increased. Every platform has now copied TikTok's model, with products such as Instagram Reels and YouTube Shorts. These short-form video platforms provide huge increases in dopamine stimulation[41] – far more than text and photos were able to prior to their invention. Given this huge, fast increase in dopamine, this is the aspect of your phone that can cause a low dopamine state to arise. You may have noticed, and certainly will observe now, that when you scroll these videos for a while, it is awesome during the scroll, but when you eventually put your phone down, you feel lethargic and low. This is your brain being zapped and depleted from the vital resource, dopamine.

Alongside the short videos, the constant notifications, emailing, messaging, scrolling stories, and seeing negative news alerts are all providing quick dopamine hits to your brain.[42] In order to improve your relationship with your phone, we are going to focus on two key aspects. Building these into your life will transform how motivated you feel each day.

Strategy

First, Phone Fasting, or taking a break from your phone. You may be familiar with fasting as a religious practice where individuals spend a prolonged period of time without food. This is something humans instinctively did as a spiritual practice for thousands of years. Interestingly, modern science is now showing the benefits this can have for our physiological and psychological health (more to come in Chapter 13, Gut Health). Just as we are mindful about not overconsuming food, we must become mindful about not overconsuming content on our phones. Phone Fasting brings this concept to your relationship with your phone. We have reached a point as a society where we must now develop a clear, daily practice whereby we 'fast' from our phones. This will enable your dopamine to replenish and open the opportunity for more connection and restoration in your life.

Secondly, how you set up your phone. We must take advantage of the various features on your phone that enable you to track your usage, alongside disconnecting you from it on a more regular basis.

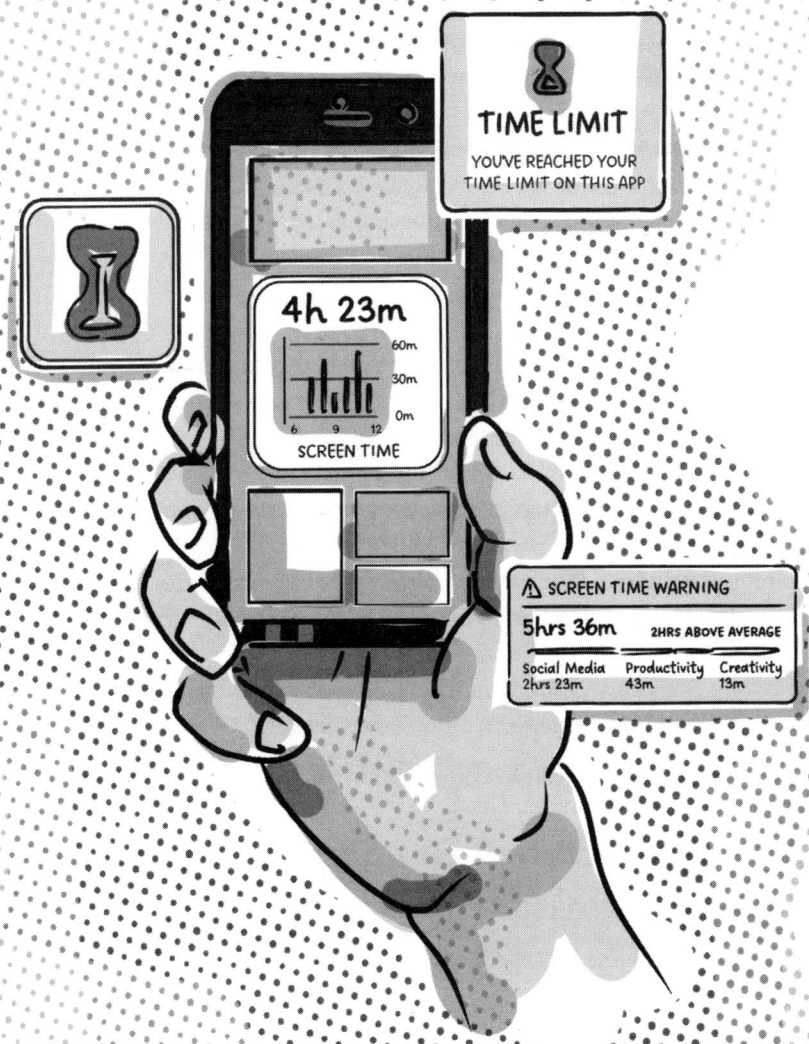

STAGE 1:
PHONE FASTING

Phone Fasting is a simple and impactful concept I developed for myself when working to manage my addiction to my phone. There is a very specific reason that I chose this route in order to alter your relationship with your phone. What you may have discovered is that if your phone is near you, no matter how hard you try, you inevitably pick it up. Say, for example, you are trying to watch a TV show and it is on the sofa next to you. No matter your resistance, if you feel a tiny bit of boredom, you find the phone is suddenly in your hand. In order to significantly reduce your phone usage, there must be windows of time in your day when your phone is not near you.

The first component of Phone Fasting involves the mighty challenge of not going on your phone when you first wake up. Now I do appreciate how challenging this is. However, after personally teaching over 50,000 people how to improve their mental health, this is without a doubt one of the most popular and impactful changes I have seen people make.

Throughout the night, as part of the restorative process of sleep, your brain is regenerating your dopamine. This makes sense, as you would want your brain to wake with an abundance of dopamine so that you feel motivated to start your day[43] (further insights to come on optimizing your Deep Sleep during Part 3, Serotonin, the primary chemical involved in regulating your sleep patterns). In this moment of waking, with a plethora of dopamine in your brain, two things can occur. One, you complete an activity that requires some form of cognitive or physical effort and your dopamine levels begin to rise. Two, you enter your phone, the world of notifications, social media, and news, and immediately you spike your dopamine system and crash it out. Remembering that dopamine is responsible for all of your drive and focus for your day ahead, it is vital we set your dopamine system off in the right direction.

Your Morning Routine

Given what we learned in Chapter 2 about Discipline, beginning your day by making your bed is a great way to start creating dopamine right away. After this, immediately go and brush your teeth and splash some cold water on your face. If you typically go to the toilet first thing in the morning, this is a time when scrolling can often occur. This was a habit that was incredibly challenging for me to break. With the knowledge that effort builds dopamine, I replaced these few minutes on the toilet with reading. As we know from Flow State, reading is a fantastic activity for your brain. If you make it a daily morning practice to read a few pages of *The DOSE Effect*, your brain will instinctively be drawn in the direction of education throughout your day. This occurs as a result of the interesting neuroscience that demonstrates wherever we receive our first dopamine hit from, our brains crave in this direction throughout our day.[44] Selecting one of these healthy dopamine behaviours first thing in your day is therefore essential.

KEY EXAMPLES of healthy morning behaviours:

1. MAKING YOUR BED
2. GOING OUTSIDE
3. COLD SHOWERS
4. WASHING FACE
5. READING
6. BRUSHING TEETH

After this there are two routes your morning can go. One is immediately getting ready for your day, having your shower and getting dressed. The other is getting outside for a short period of time to see some sunlight and potentially a natural environment. The goal is simple. Don't check your phone until either you are ready for your day, or you have been outside and seen daylight. A simple phrase I say to myself every morning in my mind is . . .

'I must see sunlight before I see social media.'

This is a good mantra to live by.

Great, now your first component of Phone Fasting is complete, well done.

Throughout your DOSE journey, additional components will be built into your morning routine to further optimize your brain chemistry.

Your Evening Routine

The second component of Phone Fasting involves how you engage with your phone in the evenings. It is incredibly easy to spend a significant amount of your evening mindlessly scrolling through the various social media platforms and continuing to check in on your work. Without noticing, time can fly by and the true purpose of your evening, whether that is connection, rest, or exercise, is lost to the world of quick dopamine hits within your phone. With this in mind, you will now make a commitment to yourself to have a prolonged period of time away from your phone each evening. Our aim is to achieve a minimum of sixty minutes without seeing your phone at all. There are four different times in your evening you could achieve your Phone Fast.

1. EXERCISING

Exercise is a perfect opportunity to have some time away from your phone. This could mean going to the gym and leaving your phone in the locker, or it could mean heading out for an evening walk, jog, cycle, or swim with your phone in airplane mode. Another brilliant way to get away from your phone in the evening is through engaging in a sport. I recently reignited my love for tennis. Sport is something I loved as a kid and building it back into my life as an adult has been incredible. Further insights to come in Part 4, Endorphins, where we will strengthen your relationship with exercise.

Please note, many people will identify with, 'But I neeeed my music when exercising'. I'd guide you to experiment with exercising totally phone-free. However, if this is not an option, ensure your phone is on airplane mode and you ONLY listen to music. Do not engage with any form of notifications or scrolling.

2. EATING

Eating dinner is a great way to spend some time disconnecting from your phone. While cooking and eating, put your phone in another room. You could still have some music, a podcast, or an audiobook playing if you need some stimulation. Our aim here is to get away from the quick dopamine and constant notifications. For those of you with families, there is incredibly interesting research demonstrating the significant value to youth mental health of involving your children in the process of cooking meals, sitting together as a family, and collectively cleaning up.[45] It helps young people to get away from their technology, connect with the family, and learn the importance of contribution. Contributing to others is a vital component of a healthy mind and something we will dive into during Chapter 11.

3. SOCIALIZING

Social connection is one of the core ingredients of a healthy and happy mind. Progressively phones have begun to interrupt the quality of social connection we can experience. I am sure you have noticed the frustrating feeling when you are trying to have a conversation with someone and they continuously check their phone. Whether you are going for a walk with someone, going out for a meal, or hanging out at your home, use this as an opportunity to spend a minimum of sixty minutes away from your phone and connect with your loved ones.

4. WATCHING TV

Now this is an unusual one, you may think, watching TV . . . what's the difference between that and scrolling? I'm sure you have found yourself struggling more and more to watch TV without continuously checking your phone throughout the show or movie you are attempting to watch. When it comes to your brain, and particularly your dopamine, when we look at how TV and social media compare, there are significant differences. Watching TV requires your brain to engage in at least some effort, as you have to concentrate and pay attention to what you are watching in order to experience some pleasure from it. Scrolling social media videos requires zero effort in order to receive pleasure. We used to worry that watching TV would destroy young people's minds; now, I'm in schools training our young people to spend some time watching TV without their phones, and that is actually a win. (Of course, the number one priority is getting them outdoors, connecting with one another.) This evening, and from now on, when you watch TV, put your phone in another room and observe the difference in how you feel.

When Phone Fasting during the evening, whether that is during a social activity or when watching TV with your family or friends, it is important the people you are with also engage in this challenge. The reason this is vital is as a result of something called anticipatory dopamine.[46] You may have noticed that when you see someone pick up their phone, you suddenly feel a quick urge to pick up yours as well. This can also happen with other quick dopamine activities. If you walk past a bar and see someone drinking a glass of wine, you may feel an urge to have one. It is the same with unhealthy food, smoking and so on. Anticipatory dopamine is where your brain observes someone engaging in a dopaminergic activity and creates a rise of dopamine in your brain in anticipation at the thought of you engaging with it, too. This rise in dopamine then creates desire for you to receive dopamine in this form as well. In order to successfully Phone Fast with others, it is essential that no one picks up their phone, otherwise resisting the urge created by this anticipatory dopamine will be incredibly challenging.

STAGE 2:
PHONE SET-UP

1. Screen time

Ensure that you have the screen time widget located either on your home screen or second screen so that you can see your usage on a daily basis. The graph below provides a simple indication as to what is healthy vs unhealthy usage. If you are at one hour per day, this is incredible. Your brain will be capable of thriving. If you are at two hours a day, this is also a great healthy number. Three hours is when you begin to hit the upper limit of what your brain will be able to manage. Above three hours, into four, five and six is when your brain will struggle. Remember, this isn't unusual if your screen time is very high. Phone Fasting to reduce your daily usage will have a huge impact and significantly improve how you feel in your mind.

2. Notifications

Ensure you have almost all of your notifications off. This may seem unusual at first, but we have got incredibly used to feeling as though we need to be pinged all the time. This spikes your dopamine system and creates much bigger cravings for your phone. Besides emergency calls from your 'favourites', turn off all notifications on all your apps. (Turn off badge app icons too.) This enables you to feel empowered to go into apps when you choose to use them, rather than feeling compelled to go into them all the time when they ping you.

3. Social media apps

Move all of your social media apps into a folder on your second screen. If your social media apps are easily accessible, it will increase the frequency with which you engage with them.

Social Media Moments
A simple daily rule for social media addicts

As someone who has a great deal of difficulty managing my addiction to social media, a rule I have created in my life that has had a transformational impact is what I call 'Social Media Moments' or 'SMMs'. These are three key moments in my day I allocate to scrolling social media, guilt free.

As we know from our understanding of dopamine, one of the greatest challenges is that if we open social media in every moment of boredom, we wear it out. With this in mind, if we select three key moments where we scroll, it will enable our dopamine system to regenerate throughout the gaps.

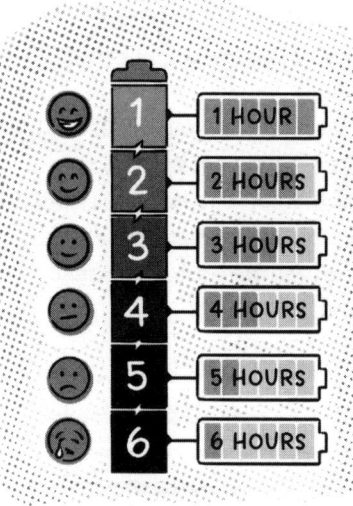

Ask yourself which app you are most addicted to. Once you have your answer, I want you to select your three moments. My three moments are **10 a.m., 3 p.m.** and **8 p.m.** I have chosen these three so I . . .

1. **Avoid social media first thing in the morning to help me remain motivated at the start of the day.**

2. **Don't spend all of my lunch break scrolling.**

3. **Get my exercise done before I end up scrolling and procrastinating.**

4. My rule for my final scroll at 8 p.m. is that I have to have eaten my dinner and cleaned it up before I am allowed on social media.

Take a moment to consider when your three moments would be in your day. There could be one in the morning, one around lunchtime/early afternoon and one after dinner. Once you have selected them, tell a friend or family member about your SMMs to commit yourself to this new idea. No matter how desperate you get, do not go on social media outside your chosen moments. Although it is hard at first, your willpower will grow quickly and the positive feeling you will experience as a result of reduced social media checking will motivate you to continue.

Often when people first integrate this SMMs rule into their lives, they observe that even without noticing their thumbs are so habitually engrained to swipe to the apps on your homescreen and click them. It is vital you click 'x' on each social media app and click 'remove from homescreen'. Then, during your SMMs you simply search for the app. People often share that this has a massively beneficial impact.

Challenge

I would now like you to complete the **'Phone Fasting Challenge'**. In order to complete this challenge, you need to Phone Fast when you wake and once every evening for the next seven days.

Throughout this challenge, aim to include others in it with you. A friend, partner, or a family member. This increases your accountability to one another alongside creating the opportunity for you all to have a discussion about your daily phone usage. Be sure to carefully observe any changes in how you feel as a result of spending less time on your phone. Check in with each other about which phone-free activities you enjoy most!

4

Embrace Discomfort

First, let's rate your ability to get out of your comfort zone.

Rate yourself on a scale from 1 to 10 for how good you are at pushing yourself outside your comfort zone. Be honest with yourself.

1 ➔ 10

1 = terrible
10 = amazing

Understanding COLD WATER

Over the last few years, I am sure you have observed the meteoric rise in the popularity of cold showers, ice baths, and sea swimming.

You may be familiar with a very interesting Dutch man, Wim Hof, who has been a true pioneer in popularizing this phenomenon. Throughout this chapter, I will provide you with the science that proves why immersing our bodies in **Cold Water** can have an incredibly positive effect, not just by supercharging your dopamine system, but on account of its host of additional physical and psychological benefits. I am, of course, very aware of the fact that freezing yourself in a cold shower doesn't sound like a particularly pleasurable or desirable experience. With this in mind, I will offer the best sales pitch I can for my belief that this is a tool worth adding into your life.

Before I begin explaining this, I want you to take your mind to a time you jumped into cold water before. Maybe a moment on a holiday when you ran into the cold ocean or jumped in a cold swimming pool. You may have hated the thought of it, and as you entered the freezing water, you may have screamed. But I am sure you experienced that refreshing, energizing feeling once your body was submerged, and the subsequent feeling of accomplishment once you got out and got yourself warm. A popular Michael McIntyre sketch that I love (Michael is an awesome comedian in the UK) describes the annoying experience when you are trying to convince yourself to jump in the cold water while your friend is swimming along in a nonchalant way, saying, 'Oh, it's fine once you're in'. As we will come to discover, that is actually true. Your body is an incredibly capable physical machine that has the capacity to survive in the wild. Once you understand the benefits of cold water, this will become a key practice supporting you in your life.

In order to understand how immersing your body in Cold Water can be of benefit to your dopamine levels, we must return to our pleasure–pain balance model (page 23). As we discovered at the beginning of Part 1, your brain's ability to experience both pleasure and pain are co-located and operate as a see-saw. During hard, painful activities, our brains evolved to evoke a pleasure response in order to keep us motivated and resilient to the challenges of the ancestral journey that brought us to this point as humans. Additionally, if during this journey we sought only for pleasure, through sex or food, with little focus on the effortful activities that enable survival, our brain would evoke a 'pain' response to reinforce that this is not the correct lifestyle for optimal survival.

Now take a moment to imagine yourself standing in a freezing cold shower. This is going to be far from a pleasurable activity; this is the opposite, pain. In this moment of pain, experienced as the Cold Water hits your body, the pleasure–pain mechanism is going to kick into action. To provide you with some context as to just how strongly it will kick into action, let me explain how certain modern-world behaviours impact dopamine levels.

In order to understand this, we must remember that at any one moment you have a certain amount of 'baseline dopamine', the amount of dopamine currently circulating through your brain and bloodstream (see page 28). In contrast, if you look at an incredibly addictive, powerful drug such as cocaine, it has been shown to increase your baseline dopamine level by two and a half times,[47] and it does so incredibly quickly by activating the pleasure response in the brain. Very soon after this, once the cocaine wears off, the alternative side of the see-saw will kick into action in order for the brain to return to a state of balance, thereby creating a proportionate 'pain' response in the brain, experienced as anxiety or depression.

Alternatively, a fascinating research study assessing the physiological response of the human brain and body to different water temperatures revealed Cold Water could also increase baseline dopamine levels by two and a half times,[48] the same as cocaine! This occurs as a result of the Cold Water evoking the 'pain' response within the brain, and therefore, during the rebalancing of the see-saw, creating a rise in pleasure and, of course, motivation, for the individual. So instead of creating a negative impact, the Cold Water creates a positive feeling. An additional reason why dopamine levels increase significantly during Cold Water immersion, and something valuable to know when applying this strategy to your life, is how dopamine and adrenaline are interconnected.

Adrenaline is the hormone responsible for energizing the body and increasing your alertness. Interestingly, dopamine and adrenaline operate as 'cousins' within your brain and body. During the moment that the Cold Water hits your body, adrenaline spikes as a result of the body perceiving there may be danger in your environment, and with this fast rise in adrenaline, dopamine rises accordingly, providing a significant increase in both motivation and focus.

Alongside this incredible rise in dopamine – one of our primary goals when applying DOSE to our lives – there are a host of additional benefits that occur. Cold Water immersion has been demonstrated to improve the functioning of your immune system.[49] One study revealed that individuals in the Netherlands who took a cold shower every day for ninety days were 29 per cent less likely to call in sick than those who didn't. Alongside this, cold showers have been seen to reduce depressive symptoms,[50] support muscle and joint pain,[51] improve exercise recovery,[52] and, incredibly, to reduce body fat through burning 'brown fat' during the period the body is cold.[53]

With all we know about dopamine and how it creates your drive, you can imagine how valuable it would be to significantly increase it at the start of your day. With the following strategy and corresponding challenge, you will be experimenting with cold showers in the mornings in order to achieve this. This will lead to improvements in focus and productivity during your morning Flow State activities.

When I first came across all this research, I was genuinely almost annoyed. I didn't like the idea of having Cold Water hit my body in the mornings, especially in the winter. Over a few months I experimented with a range of strategies in order to introduce this to my life in a sustainable way. The following strategy will enable you to achieve the same.

> IMPORTANT NOTE: If you are experiencing high levels of anxiety, stress, or loss of breath to a point where you struggle to recover and return to your 'normal state', then I strongly advise avoiding engaging with Cold Water therapy. Additionally, when battling cold or flu viruses, it is best to avoid taking cold showers. If you are navigating a specific health condition and are unsure about how Cold Water immersion could impact it, be sure to chat with your doctor first. During these moments, 'Heat' is your friend, as you will come to learn in Part 4, Endorphins. Please always consult your GP.

Remember, developing the ability to overcome the part of your mind that doesn't want to do challenging things is essential to incorporating healthier behaviours into your life! Cold Water immersion is exactly how you can achieve this.

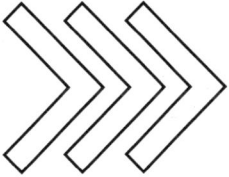

Strategy

In order to create a sustainable habit with Cold Water immersion, I don't want you to get in a freezing cold shower, hate it, hate me, and throw this book out the window. Our aim is to gradually ease into cold showers. Tomorrow morning, get in the shower when it is warm, and spend your usual time washing yourself. Once that's done our aim will be to turn the shower cold; not so cold that you do actually hate me, just cold enough that you feel a 'pain' response in your brain and body. During this moment, begin taking deep, slow breaths in, and long, full exhales out. Remember, the aim isn't to enjoy it. Anyone who tells you they love it is weird – this is supposed to be hard, and the difficulty is what supercharges your dopamine and trains your resilience. These slow breaths will calm your nervous system (more to come on understanding your nervous system in Part 3, Serotonin). Once you have spent five to ten seconds under the Cold Water, turn the shower off and get out. Finishing with cold, rather than returning to hot, is important as this period of your body reheating will contribute to your dopamine production.

The next morning, our aim is to simply turn the shower a little colder and stand in there for a little longer. Benefits will occur from the moment that you experience any Cold Water on your body, but we are striving to reach a point where you spend thirty to sixty seconds in cold running water. An interesting additional piece of guidance is that if you can ensure your head or face are under the water when you turn it cold, this will make it easier. It sounds counterintuitive but there is some fascinating neuroscience known as the 'mammalian dive response', which shows when our heads are submerged in the cold our body will actually regulate its temperature faster.[54] This is why it's often easier to jump into the ocean rather than slowly walk into it.

MUSIC
An additional tip that has made a HUGE difference for me has been through utilizing music. Music is incredibly powerful for a great range of reasons, as we will discover in Part 4. Put a song on before getting in the shower. Then, when the song gets to the chorus, or the part you like the most, turn it cold at this moment. Sing, dance, move your body. The music will elevate your mood and motivate you to stay in the Cold Water for longer.

> **Hangover TIP**
>
> As we know from our progressive understanding of dopamine, drinking alcohol is one of the most common behaviours many of us engage in that crashes out our dopamine system. As Cold Water provides the opposite effect, utilizing it during the days following any drinking will be valuable. A simple way you can achieve this is by filling a large bowl with water and ice and then dunking your face in it. Complete three rounds, holding your breath each time for ten seconds. Of course, be careful; don't hold your breath for too long or give yourself too much brain freeze! If this feels too far out of reach, remember that simply splashing Cold Water on your face in the bathroom sink will provide a slight dopamine increase, something I highly recommend doing every morning immediately after waking.

Challenge

I would now like you to complete the '**Cold Water Challenge**'. In order to complete this challenge, you need to turn the shower cold every morning for the next seven days.

Throughout this challenge, aim to convince a friend, partner, or a family member to do it with you! This increases your accountability to one another, alongside creating the opportunity for you all to have a discussion about the challenges and subsequent benefits that occur when engaging in Cold Water therapy.

5

Find Your Goal in Life

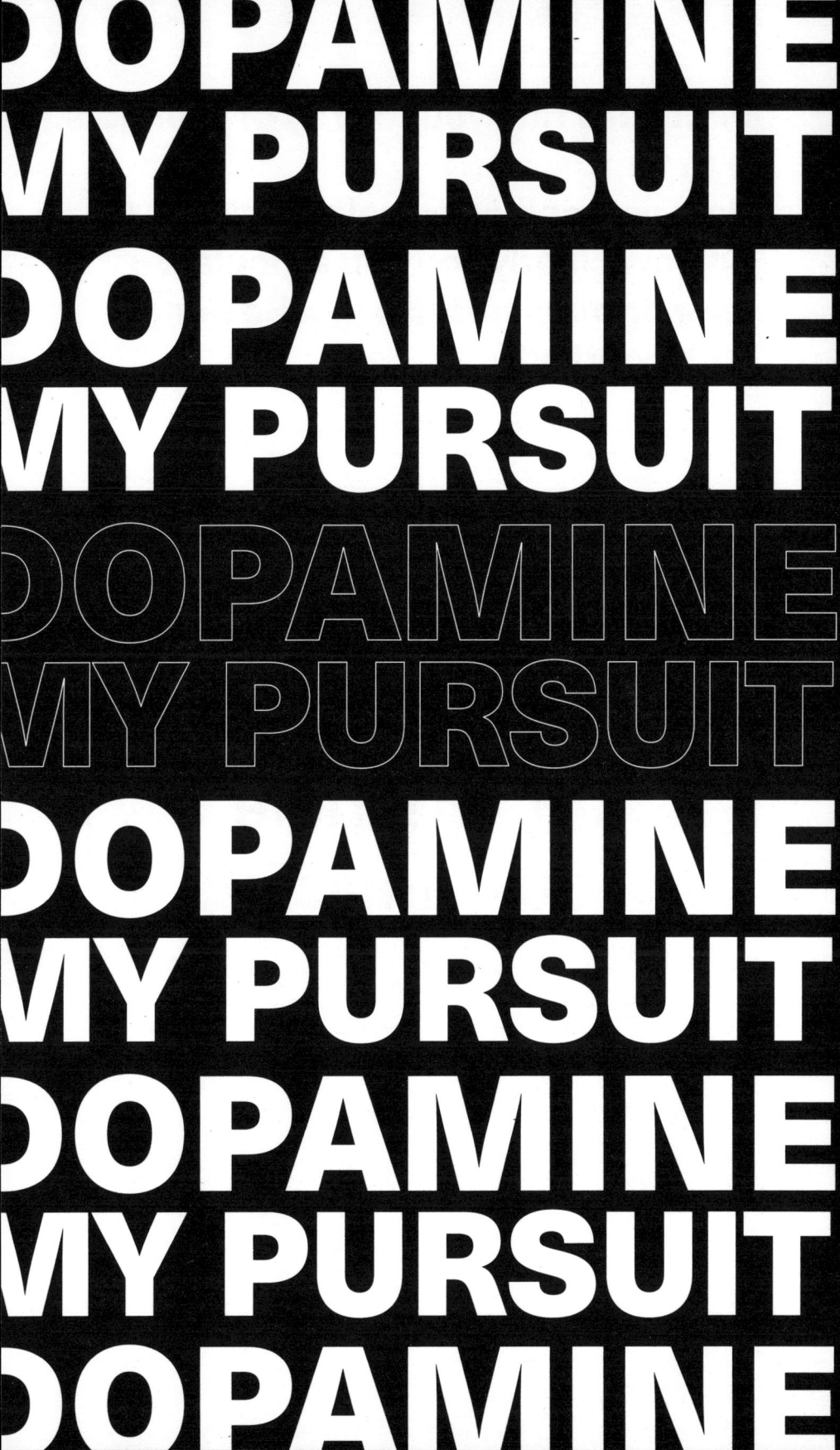

First, let's rate your clarity of your goal.

Rate yourself on a scale from 1 to 10 on how clearly you can articulate your primary goals in life. Be honest with yourself.

1 → 10

1 = terrible
10 = amazing

Understanding MY PURSUIT

We have arrived at the final chapter in Part 1. A chapter that is, from my perspective, the most important component of dopamine and potentially in the whole of *The DOSE Effect*. It is going to get you thinking about what you are really seeking in your life – and, importantly, what you are willing to sacrifice in order to attain it.

For many years, dopamine has been known as the 'reward chemical', the chemical that provides us with a pleasurable feeling when we accomplish something. This doesn't capture the true function of dopamine. As we know from our hunter-gatherer learnings, dopamine would rise in the pursuit of a goal, be that hunting for an animal, building shelter or foraging for food. Dopamine would rise to provide us with the required level of motivation and deep focus to successfully complete what we were seeking to achieve. We believe our aim in life is to accomplish our goals, but the best feeling actually comes from pursuing them.[55]

When observing individuals who win the lottery, or celebrities who achieve astronomical success, or medal winners, you see this science in action. When this success is experienced, often mental health challenges can follow. This can occur as a result of individuals no longer having anything to pursue or chase. It is unbelievably important to our neuro-biology that we have a need for progress. If we didn't have this deep drive, as hunter-gatherers we simply wouldn't have survived. If, for example, I came across a hunter-gatherer tribe and provided them with all the resources they could ever imagine, initially they may appear excited, but over a number of months they would show demotivated, depressive, lazy symptoms. They would no longer be engaging in those hard activities that build slow dopamine, and would begin to suffer the consequences.

It is incredibly important you have a clear pursuit in your life, one that feels possible to accomplish but is simultaneously challenging and will require a significant amount of effort. Not only does this build your dopamine, but having a specific goal in mind supports managing the variety of addictive behaviours we engage with. As I mentioned at the start of this book, I have found learning to manage a wide range of quick dopamine behaviours incredibly challenging. Over the last three years, once I became very clear on what **My Pursuit** is, my relationship with these behaviours changed significantly. Now that I have a true purpose, when I consider drinking too much alcohol and being hungover, or watching porn and killing my drive, or eating badly, or overscrolling my phone and procrastinating, I come back to My Pursuit and think, okay, am I willing to sacrifice these quick hits of dopamine for the long-term sustained happiness that this pursuit is providing? As we go on to decipher what your pursuit is, this will help you to sacrifice the short-term hits of pleasure for this longer-term form of true joy.

An additional reason why it's vital to have an incredibly clear goal in your life is seen when you go deeper into the true function of dopamine. Dopamine is designed to keep us alive, and achieves this through keeping us engaged in the activities that increase its likelihood. In order for your dopamine levels to remain at the healthiest, highest levels they can, you need to have an excited anticipation about your future,[56] just as a hunter-gatherer needed to know they had food and shelter. However, in our modern lives, this mechanism is a little more complex. Take some time to consider a couple of moments in your life. One when you have felt incredibly low, and one when you have felt incredibly happy. Often in low moments, our future may seem bleak and not like the one that we would like. This can occur as the result of losing a job, losing a partner, health challenges, or any array of experiences that take away your excitement about your future. On the other hand, in extremely happy moments, maybe a job promotion, the feeling of falling in love with someone, or booking yourself an exciting new holiday away, you experience excited anticipation about what's to come in your life.

There is a wide range of areas of human experience that can provide you with a goal to pursue. We are going to decipher which feels the most appropriate for you in your life right now.

In order to do this, we have five areas you could consider as your 'pursuit': career, family, health, creativity, and your DOSE. After you have read the five areas, I will provide you with a simple exercise to complete in order to identify your primary pursuit.

1. CAREER

Our careers are one of the most important aspects of our lives. We spend a huge amount of our time working. Considering what your primary pursuit is within your career is a valuable exercise. You may have noticed a really positive feeling arising in moments where you know you are making progress within your career. Maybe you've excelled on a project, worked towards a promotion, or been celebrated for your contributions by a colleague. All of these experiences create a significant rise in your dopamine levels as you work towards your given goal.

Take a moment to ponder what you want most in your career right now. Is there a project you are wanting to nail? Is it that you are seeking a specific promotion or pay rise? Are you wanting to take on more responsibility in your position?

Selecting a very specific goal you deeply seek to attain and working towards it on a regular basis will support you to optimize your dopamine levels alongside, of course, improving your life.

2. FAMILY

Our family is a vital component of what makes us human and what creates true happiness. We will go on to deeply discover the importance of your family relationships during Part 2, Oxytocin. I, of course, recognize that everyone reading this book may be in a slightly different position. It may be that you are in your late teens, your twenties, your thirties, potentially raising young children, or your forties, fifties, or sixties, where you are caring for your family.

The pursuit of deepening your familial connections is a brilliant way to optimize your dopamine alongside building your oxytocin. Take a moment to ponder how you could pursue strengthening your family bonds. Is it that you feel you should reconnect with one of your parents? Spend more time with your siblings? Take some time out to connect with your grandparents? Or spend more time phone-free, deeply connecting with your children? Or maybe reconnect or heal a friendship with someone you have become disconnected from in recent times.

Selecting a specific family-based goal, and pursuing it with effort and attention, will support boosting your dopamine levels, alongside creating more love in your life.

3. HEALTH

Your health, as we are discovering and will learn more and more deeply throughout *The DOSE Effect*, is one of the most important aspects of your human experience. A friend of mine actually said to me recently, 'I want to get my life in order and it all starts with my health.' This, in my opinion, is an incredibly accurate statement.

As we journey into serotonin and endorphins, you will learn a great deal about how you can optimize your health through Gut Health, Nature, Exercise, Deep Sleep, and much more. When considering this concept of the pursuit of a goal, your health is a perfect place to start. You may have noticed before that when you start eating healthily, or working out more, it makes you feel wonderful. It is important to understand this feeling occurs both as a result of the physiological benefits of the nutrition or movement, but also as a result of the progressive accomplishment towards a goal.

Take a moment now to consider what your primary health goal is in your life. Is it that you feel you need to eat more nourishing foods? Is it that you need to get out of your home more and move your body more often? Is it that you need to prioritize the quality of your sleep? Do you finally need to make a doctor's appointment to get your worry checked out?

The pursuit of this health goal will optimize your dopamine levels alongside providing a host of additional health benefits associated with these health-focused behaviours.

4. CREATIVITY

Humans innately have incredibly creative brains. It is actually an aspect of us that is progressively getting harnessed less and less as we move towards a more technologically focused world. Where free time once might have been spent drawing, painting, reading, or writing, we often now find ourselves sitting on the sofa in front of the TV with our phones in our hands. Well . . . not any more now that you're becoming a pro Phone Faster!

If, for example, you choose to complete a puzzle in your free time, or begin some form of art or musical pursuit, it will have an incredibly positive impact on your life. Your brain deeply wants to make progress towards goals, and creative outlets are a perfect way to achieve this.

What could you pursue as a creative outlet in your life? Could you do something creative with your home, some form of DIY project? Could you learn to play a musical instrument? Could you find an artistic project? There are incredible 'adult colouring books' that I've found so calming., or, you may find the pursuit of reading this book is providing a creative outlet.

Take some time to ponder what your creative pursuit could be. There are a host of additional benefits that will arise as a result of this. It may be that you could do this activity with someone, a friend, your kids, your partner, or your siblings. Alongside this it will further support your Phone Fasting goals of spending time away from your technology and social media, allowing your dopamine to recharge simultaneously.

5. YOUR DOSE

I always love explaining the pursuit aspect of dopamine, as now you can deeply understand why I have constructed this book in the way I have. The entire reason I made this book formulaic, with strategies and challenges for you to complete, is that immediately through picking it up you have already entered the pursuit of a goal. Whether it has been through improving your ability to get into deep Flow States, being more Disciplined in your home environment, Phone Fasting on a regular basis, or freezing your ass off in Cold Water, you are now in the pursuit of accomplishment.

If you are unsure what your pursuit should be, I've got you covered. Your DOSE is your pursuit for the period of time you are reading this book and throughout your life going forward.

The simple concept of understanding that your brain deeply, biologically, craves accomplishment, and finding your unique way to create this feeling, must become a priority in your life.

Strategy

In order to identify your pursuit, I am going to suggest one of, if not the, most impactful practices I have ever added to my life: the phone-free morning walk. I genuinely cannot even put into words what this has done to my life, my career, and my mental health.

Take a moment to consider which of these five stands out to you as the pursuit you want most in your life right now.

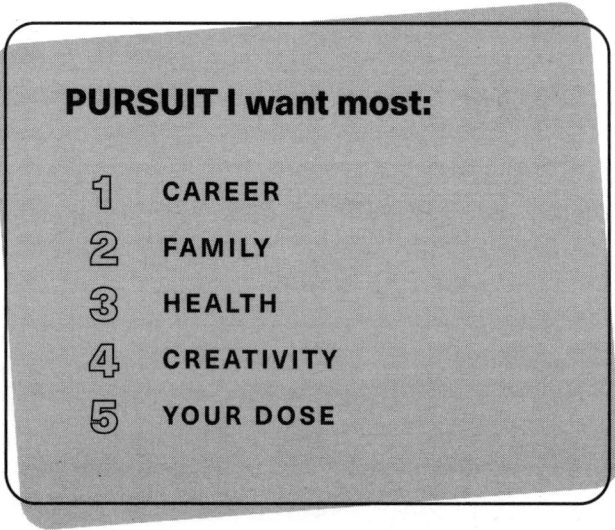

PURSUIT I want most:

1. CAREER
2. FAMILY
3. HEALTH
4. CREATIVITY
5. YOUR DOSE

Tomorrow morning, I want you to go on a walk in nature, without your phone. If you need your phone in order to feel safe, then please put it in your bag on airplane mode. Take yourself to a nice natural environment and ask yourself the questions, 'What is My Pursuit?', 'What am I really seeking for in my life right now?' Too often in our modern world we have hopes and dreams but remain distracted by the quick dopamine that constantly draws our attention away from them.

If you spend forty-five minutes or so deeply pondering this question and chatting away to yourself, you will come home with a vision and a plan in order to begin this mission. I was someone who never ever used to spend time in silence. I was once at university, working in the library when my phone ran out of charge. I had a fifteen-minute walk home from the library and hated the idea of fifteen minutes of silence. It sounds ridiculous when I write it now, but I actually carried my laptop in my hand with my headphones plugged into it in order to avoid this silence. I deeply relate to the desire for distraction, always listening to podcasts, music, or TV shows.

I do, now, however, believe that the time I have spent in silence over the last few years is the single most impactful practice I have added into my life. During our chapter on nature, we will go deeper into this topic of silence and how I can get you not just to feel comfortable with the quiet, but begin to prioritize it as one of the most important aspects of your life.

Challenge

I would now like you to complete the **'My Pursuit Challenge'**. In order to complete this challenge, you need to go for a walk in nature, in the quiet, without your phone, and ponder the number one goal you are seeking in your life right now.

Throughout this challenge, ensure you discuss your pursuit with someone you are connected with. Open the conversation by describing to them specifically what you are aiming for in your life right now. Offer them the opportunity to consider the same.

Building Your DOSE

Throughout Part 1, you have discovered the importance of understanding the true power of your dopamine system. You have been on an awesome adventure of experimenting with a range of new behaviours that will support optimizing this system in your brain.

I now want you to take some time to consider which one of the five primary dopamine actions feels the most important for you to continue to prioritize. It would be incredible if all of these behaviours remain a priority in your life. However, selecting one core behaviour and ensuring it becomes deeply embedded in your life is essential. Take this one step at a time and celebrate every positive change you make to your habits.

WHAT WILL BE YOUR PRIMARY
Dopamine Action?

1. FLOW STATE
The ability to enter deep states of focus.

2. DISCIPLINE
Maintaining an organized, clean, and calm home environment.

3. PHONE FASTING
A daily commitment to finding space away from your phone in the mornings and evenings.

4. COLD WATER
Turning your shower cold on a daily basis to supercharge your motivation.

5. MY PURSUIT
The consideration and selection of the primary goal in your life you are seeking to attain.

Key understanding: the most important principle to take away from your incredible understanding of dopamine is that you are in control of how motivated you feel. You are in the driving seat. If you succumb to the temptation of the quick dopamine hits too regularly, you will undoubtedly feel lethargic and unmotivated. Pursuing anything meaningful to you will feel like a chore. If you, however, prioritize these new dopamine actions you have learned, your dopamine levels will rise, and pursuing the life you dream of will become easier than ever before.

Be sure to chat with a friend or family member about the primary dopamine challenge you have selected!

PART 2

Feel Confident and Connected

OXYTOCIN
OXYTOCIN
OXYTOCIN
OXYTOCIN
OXYTOCIN
OXYTOCIN
OXYTOCIN
OXYTOCIN
OXYTOCIN
OXYTOCIN
OXYTOCIN

Understanding OXYTOCIN

Welcome to Part 2 of your DOSE journey, Oxytocin. Oxytocin, described as 'The Great Facilitator of Life', is one of the most powerful chemicals in your body and one of the most vital to both our procreation and our survival as humans.[1]

This is because of its powerful impact on our desire as humans to connect with one another, form bonds, work closely in groups, and have children. For many years I have been on a huge journey studying this chemical, learning how I can boost it in my life and the lives of the people around me. This is because I have seen the transformational impact oxytocin can have on relationships, self-belief, and the way in which we view our world.

Throughout Part 2, we will be diving into the true function of oxytocin, looking at how our modern lifestyle can lead to a reduction in the activation of it, and, most importantly, you will complete a range of challenges in order to learn how to boost your oxytocin levels.

To understand oxytocin, we have to journey back to the moment you were born. A moment that you don't remember, but the moment when oxytocin first surged through your brain and body.[2] In the moment of childbirth, oxytocin is released in your mum's brain as well as your brain in order to create the initial pair bonding between you both.[3] For your mum, it creates a strong, innate desire to deeply care for you, to give you love, to protect you, and to keep you safe. For you, as a newborn baby, it creates a deep wanting for that love, for that connection from your mum and the caregivers around you, which will help keep you alive.[4]

The Oxytocin Principles

PRINCIPLE 1:
REQUIRES GOOD-QUALITY IN-PERSON SOCIAL CONNECTION

Throughout the first few months of your life, you begin receiving love in many forms, such as through breastfeeding, physical touch, or loving words This experience progressively builds more and more oxytocin within you, creating a deep bond between you and those who are caring for you.[5] As you make your way through your life, oxytocin remains incredibly important. During any moment in which you are receiving love or giving love to others, both you and the person you are connecting with will experience a rise in oxytocin.[6]

PRINCIPLE 2:
REQUIRES POSITIVE, GRATEFUL, INTERNAL SELF-TALK

There is an additional component of oxytocin that is essential to understand, particularly given the impact of our digitally based lifestyles. Giving and receiving love is not just something you do with others, but something you can do with yourself as well. Our comparison-based society can often lead many of us to struggle with the inner narrator in our minds. It may be that you have an internal critic who constantly judges you on how you live your life, how you look, or how successful you are. This is detrimental to oxytocin. Instead, to optimize your oxytocin, it is essential we integrate key actions into your life that are going to quieten that critical voice in your mind and greatly improve not only the relationship you have with yourself but also your capacity to give yourself love.[7] I have had to do a lot of work to change my own negative thinking and thought patterns. I am someone who struggled deeply with a critical and judgemental inner voice. I have therefore experimented with a range of strategies and found some that have genuinely transformed that voice in my head and have led to a more loving and compassionate experience in my mind. An interesting side-effect that occurred as a result of this alteration in my relationship with myself is that it has greatly impacted my success in the various pursuits that I deem important. I feel incredibly confident that you will experience the same as you step into the next phase of your DOSE journey.

First, let's go back and consider our ancestors, and the essential role oxytocin played in our evolution.[8] As we know from our insights into dopamine, survival is at the forefront of all of our behavioural decisions. Early in our evolution, a human, on their own, would struggle to survive deep within the forest. However, an individual who works closely with a group of other humans to find food and shelter fares much better. We therefore developed a deep desire to belong to a group, as it was essential to our survival. In groups we could prosper; alone we would diminish. Consider this when looking at your life. You may have experienced moments when you have felt excluded from a group of friends, an event, a party, and felt unease or anxiety about it. The feeling of exclusion is wired within us to create fear, to motivate us back to the group. Deep within your being, you know that being alone is not an optimal path to survive and thrive.

Now, with the development of our technology-based lives, things are changing fast. Consider the difference between us once spending all of our time closely connected, working together, in person, and towards a common goal. Now think of how many of us spend a large proportion of our day behind our computers and working predominantly alone. We then often find ourselves spending our evenings behind our phones and TV screens, again not connecting with one another. This new way in which we live as a society is evidently leading to a reduction in this key bonding chemical. Exploring the ways in which we can reintegrate activities that bring us together, while continuing to live a modern, digital life, is our goal.

All of us are different. Some of you, the more extroverted by nature, may experience increases in oxytocin during large social gatherings.[9] Others, who are more introverted by nature, may get the greatest rise in oxytocin as a result of more intimate one-on-one conversations and experiences.[10] Either way, both increase your oxytocin levels and are invaluable to our wellbeing. One thing we know for certain is that we need to connect with others to feel the benefit. We need to feel loved and we need to give love to one another. Recent research revealed that more than three in five Americans are experiencing loneliness and feel as if they lack companionship.[11] To put into perspective just how significant an issue this is, the National Institute on Aging in the US considers prolonged social isolation to have as much negative impact on your health as smoking fifteen cigarettes a day,[12] and loneliness has been estimated to shorten a person's lifespan by as much as fifteen years.[13]

We need one another's love, now, more than ever. Throughout Part 2, we are going to embark on tasks that will encourage the experience of love, contribution, and connection with the people around you.

Do you have low oxytocin levels?

When trying to identify low levels of oxytocin, there are two symptoms that you might see.

First, you may be experiencing feelings of loneliness and isolation.[14] This is something I have navigated at different times in my life, and which has been particularly prominent throughout the last few years. In my twenties, I came to realize that the lifestyle I wanted to live was slightly different to that of my closest friends. For context, I grew up in a town close to London in the United Kingdom, and have always been a part of a friendship group whose social lives are oriented around a strong drinking culture. Partying is something I loved for many years of my life, to the point where my identity was closely tied to being someone who could party 'hard'. As I moved deeper into my neuroscience research and understanding of mental health, I came to the conclusion that I wanted a different life. A calmer, healthier life that leaves me feeling more energized and motivated to pursue the goals I seek most. As my desires changed, I moved away from London and this party lifestyle, which initially created feelings of loneliness and isolation. However, I put in the time, and I have been through an amazing process of rebuilding an unbelievably strong feeling of connection with my family, my friends, and my local community, and will be guiding you to do the same using a number of really useful challenges.

The second feeling that can arise as a result of low levels of oxytocin is a lack of confidence and self-belief.[15] We are living in an incredibly unusual time due to our social media-based society, and low levels of confidence are something I see and help people with every day. This may manifest as having a critical, doubtful voice in your mind that doesn't believe you are capable of achieving your goals, or it may manifest in criticism around your appearance and lack of belief in yourself in social settings. In a similar way to the process I have been on to rebuild my feeling of love and connection in my life, I have worked hard to build my confidence and belief in myself, and will show you the techniques you can use to do the same. Through Chapters 9 and 10, Gratitude and Achievements, we will be experimenting with some awesome, simple strategies that have transformed my self-belief and self-talk, and will enable you to experience the same.

The FOUR CAUSES of low oxytocin

For simplicity, I have broken the main causes of low oxytocin into four key categories. When we were looking at the reasons behind low dopamine, we were approaching things through a slightly different lens. We have created sophisticated methods to 'hack' the creation of dopamine with social media, junk food, alcohol, and so on,[16] leading to it surging and then crashing out. With oxytocin, serotonin, and endorphins, we haven't created such methods to hack these chemicals. Instead, we need to look at how our daily behaviours are leading to a suppression of their production.

1 The first of the four causes of low oxytocin is a lack of social connection.[17] This is simple, but right at the core of the challenge here. As society becomes more digital in focus, the quality of in-person interaction is reducing at pace. We will be exploring how you connect with the people in your life in Chapter 8, Social Life.

2 The second cause is the use of phones during moments of social connection. If we are constantly having our attention pulled away from those we are with in order to check our notifications, it will lead to a reduction in the quality of the connection that can be achieved. Have you ever found yourself telling someone a story, or talking about how your day went, but they are looking at their phone and ask, 'What did you say?' when you finish speaking? It's something that is incredibly frustrating and downheartening, and over time weakens the strength of our emotional connections with one another. This is an additional reason that Phone Fasting is such an essential practice in the evenings, not just to recharge your dopamine but to enable the greatest form of connection, and allow for maximum oxytocin release.[18]

3 The third and fourth causes are looking at the other side of oxytocin we are learning about – your relationship with yourself. The third cause is the detrimental impact of online comparison, leading to a critical inner voice.[19] Online comparison is without doubt one of the greatest challenges of the modern world. It causes a huge lack of recognition for what we achieve and a lack of gratitude for the lives we get to live. While social media can be wonderful, as it allows us to dream of amazing things, we also have to be incredibly careful of the expectations we put on our lives, our relationships, our careers, our appearance, and so much more as a result of looking at the carefully curated and perfected content that people put online.

A brilliant equation to keep in mind is:

HAPPINESS
is equal to
reality
minus
expectations

Read that again, slowly.

Your happiness is deeply interconnected with the expectations you have. Imagine you have a birthday coming up, and you want it to be the most amazing day of your life. If the birthday doesn't go quite as you expected, you are left with a feeling of dissatisfaction and disappointment. Alternatively, if you maintain a normal, natural level of expectation, you may find that your birthday actually exceeds what you were hoping for, and this is when the greatest levels of happiness can occur. This is because the reality exceeded your expectations. During the chapter on Gratitude, we will go through a practical strategy that will transform your relationship with comparison and keep you focused on your life and the lives of those you love.

4 The fourth and final cause of low oxytocin is self-criticism.[20] A particular focus here, in our image-conscious world, is the analysis of your appearance and the impact this has on your self-esteem and mental health.[21] This is a huge one for many of us. The world we now live in is unbelievably image-conscious. Whether you experience challenges with critiquing your body, your hair, your face, or whatever it may be, this internal conversation is incredibly harmful to your mind and your relationship with yourself. DOSE provides two solutions to this. One, learning to love yourself for who you are, and two, living your life in a way that enables you to feel truly healthy in your body.

How high oxytocin levels will feel

When your oxytocin levels are high, you will feel far more connected to the important people in your life, along with experiencing a greater level of self-belief and confidence.[22] We will now begin exploring the five key actions you can take to optimize your oxytocin and continue your journey of neuroscience-backed self-exploration. A happier, more connected life awaits.

On the next page, you will find a summary of the key functions, principles, feelings, and behaviours that are associated with optimizing your oxytocin.

Oxytocin Summary

Function →
- Relationships
- Confidence

Principles →
- Requires good-quality in-person social connection
- Requires positive, grateful internal self-talk

Low Oxytocin Symptoms →
- Lonely
- Unconfident

Low Oxytocin Causes →
- Lack of socializing
- Phones during social activities
- Online comparison
- Critical self-talk

High Oxytocin Symptoms →
- Connected
- Confidence

Oxytocin Actions →
- Contribution
- Touch
- Social Life
- Gratitude
- Achievements

6

Prioritize Other People

OXYTOCIN
CONTRIBUTION
OXYTOCIN
CONTRIBUTION
OXYTOCIN
CONTRIBUTION
OXYTOCIN
CONTRIBUTION
OXYTOCIN
CONTRIBUTION
OXYTOCIN

First, let's rate your current level of contribution.

Rate yourself on a scale from 1 to 10 for how much you feel you are contributing to others. Be honest with yourself.

1 ➔ 10

1 = no contribution at all
10 = incredible contribution

Understanding CONTRIBUTION

This chapter discusses one of the most beautiful sides to our human nature.

Have you ever noticed that when you do something for someone else, be it a friend, a colleague, or a family member, it creates a warm, calm feeling inside you? It is in human nature to love and support the people around you. Imagine your ancestors, surviving together through all of the challenges life was throwing their way. It was absolutely essential that they had a deep, innate desire to help the group in any way that they could. This feeling of group contribution has progressively reduced as we have moved away from community-focused living and towards more self-focused goals. During this chapter, we are going to discover precisely how you can contribute to the people in your life in an effective and valuable way.

The most simple example of how contribution feels can be seen on Christmas Day. As a young child, I would be unbelievably excited about Christmas, desperate to run downstairs and open presents. As I matured, I progressively found the experience of giving presents to others to be more enjoyable than receiving them. I am sure you have felt this too, the excitement of handing a gift over to someone you love, particularly if you feel that the gift is thoughtful and will be of value to them. This is contribution. This is oxytocin.

Modern neuroscience has validated our instinctive feelings that pro-social behaviours such as group support, empathy, and cooperation lead to a huge rise in oxytocin.[23] Acting selflessly and focusing on others is a necessary and admirable trait we must develop. Those of you who have watched the TV show *Friends* may remember an episode that perfectly depicts this mechanism in action. In this episode, Joey says to his friend, Phoebe, that 'there is no such thing as a selfless good deed'. He suggests that it is impossible to do something for someone else and not feel good afterwards. During the episode, Phoebe completes various activities for others, such as raking the leaves in someone's garden without them knowing, all with the intention of seeing if she can do something to contribute to someone else without feeling good herself afterwards. Phoebe discovers that it is impossible; every time she helps others, she feels good herself. This is a powerful discovery and something that is so beautiful about our inner nature. Supporting others will, of course, serve the person you help, but it will also make you feel amazing as well. This is oxytocin, prioritizing love for the group around you.

During the time I began discovering and understanding the neuroscience behind **Contribution**, I started assessing my own life and actions. When pondering my actions, I came to realize that a lot of my time was spent oriented towards helping myself and not necessarily supporting my family, my friends, and the community around me. I have therefore progressively gone on an unbelievably enjoyable and satisfying journey of orienting my daily behaviour towards supporting those I love and the people of our world. As I sit and write this book each day, for example, I, of course, get personal satisfaction through the accomplishment of the writing. However, the most powerful driving force within me is the thought of you, reading this book, in this moment, and beginning to take positive action on various aspects of your life.

As with all aspects of your DOSE journey, I have simplified Contribution into a variety of key actions and challenges for you to complete to support others in your life, alongside, of course, boosting your oxytocin levels. We will now explore four key groups you could contribute to. Read them carefully and observe which of the four instinctively feels the most important and valuable for you and those around you right now, and start there.

1. FRIENDS AND FAMILY

First, let's think about your friends and family. Take a moment to consider some of the following examples of how you contribute to those you love.

- **Financial support:** contributing to the financial welfare of your family.
- **Cleaning and organization:** contributing to cleanliness of your family's home.
- **Childcare:** contributing through support and by looking after your own, your friends', or your family's children.
- **Emotional support:** contributing by being someone your family and friends can come to in times of emotional need.
- **Cooking:** contributing through shopping and cooking meals for your friends and family.
- **Educating:** contributing through educating those you care for.
- **Quality time:** contributing through prioritizing your friends and family in your schedule and taking time to connect with them.

- **Celebrating achievements:** contributing through recognizing and celebrating a person's achievements (more to come on this in Chapter 10, Achievements).
- **Surprises:** this doesn't necessarily mean some huge surprise birthday (it can, of course, if that's what you're called towards). It might simply mean arranging a special date night for you and your partner, or it might mean cleaning your home before they get back from work, or it might mean getting someone you love a small gift they would value (this book would be a good option for that one). Surprises show that you are thinking of them, that you value them, and that you are willing to put in effort to make them feel special.

2. WORK

Your work is another area of your life where you will be making a Contribution. Consider which of the following examples resonate regarding how you are contributing in your working life.

- Delivering high-quality work.
- Creating a good team environment.
- Taking the initiative on projects.
- Being a good leader.
- Solving problems and resolving conflicts.
- Using your work to make an impact on the world.

It is easy to put in a great deal of effort to our work but not necessarily recognize our own Contribution. Take a moment to celebrate yourself for how you are supporting those you work with, and the impact your work is having.

3. CHARITABLE WORK

Doing some form of charitable work is another great way to contribute. It will, of course, provide an oxytocin release and make you feel awesome, but the most important factor is serving others in some way.[24] Here are some examples of how you could contribute in a charitable way to the world around you.

- **Volunteer at a local food bank:** Spend time helping to prepare and serve meals to individuals in need.
- **Donate blood:** Participate in blood donation drives to provide life-saving blood for those in medical need.

- **Mentorship programmes:** Volunteer as a mentor for young people or individuals seeking guidance in their education or career.
- **Animal shelter volunteer:** Offer your time and assistance at an animal shelter by walking dogs, cleaning cages, or helping with adoptions.
- **Environmental clean-up:** Participate in community clean-up initiatives, such as beach or park litter picks, to promote environmental conservation.
- **Elderly care:** Visit or provide companionship to senior citizens in nursing homes or assisted living facilities who may be lonely or in need of social interaction.
- **Fundraising for charities:** Organize or participate in fundraising events such as charity runs, bake sales, or auctions, to raise money for non-profit organizations.
- **Donate possessions:** As we know, the process of a deep clean and reorganization is amazing for your dopamine. Combine this with donating a number of your possessions to charity to get an oxytocin release.

4. COMMUNITY

Smile at strangers! Now admittedly this one does sound unusual and kind of creepy. However, this is a big one, and one that I really want you to action in your day-to-day life. In 2014, a brilliant psychologist called Gillian Sandstrom made a fascinating discovery on the power of fleeting social connections.[25] Gillian discovered that brief moments of connection with strangers can have an incredibly positive impact on our wellbeing, alongside our sense of community.

You may have noticed that if you smile at a stranger on a dog walk, it provides a small dose of joy, or if you ask the barista in a coffee shop, 'How is your day going?', you feel a sense of connection with that person. Our modern world is getting faster and faster, and many of us are operating solely in our own worlds. An amazing way to contribute to the world around you is simply to take part in it, interacting with the people you see: smile, nod, say hello. Our digital world can create a great deal of loneliness for many; that quick conversation with the assistant in the supermarket may have a greater impact than you realize.

You now have a clear understanding of a range of ways in which you could be contributing to the world around you. Whether that's in your family relationships, or with your friends, your colleagues, or your local community, it is essential that you make this a priority. We feel our best when we prioritize others.

Strategy

Consider how you would like to CONTRIBUTE throughout this next week to:

1. YOUR FAMILY AND FRIENDS
2. YOUR WORK COLLEAGUES
3. A CHARITABLE ORGANIZATION
4. YOUR LOCAL COMMUNITY

Challenge

I would now like you to complete the **'Contribution Challenge'**. In order to complete this challenge, you need to perform one random act of kindness every day for the next seven days. This could be from any of the four key Contribution areas.

Try one of these random acts of kindness:

- Cook someone a nice meal
- Help somebody with their kids
- Spend time listening to someone
- Help a colleague with a challenge at work
- Give your time or possessions to a charity
- Think of your own act of kindness

7

Make Physical Connections

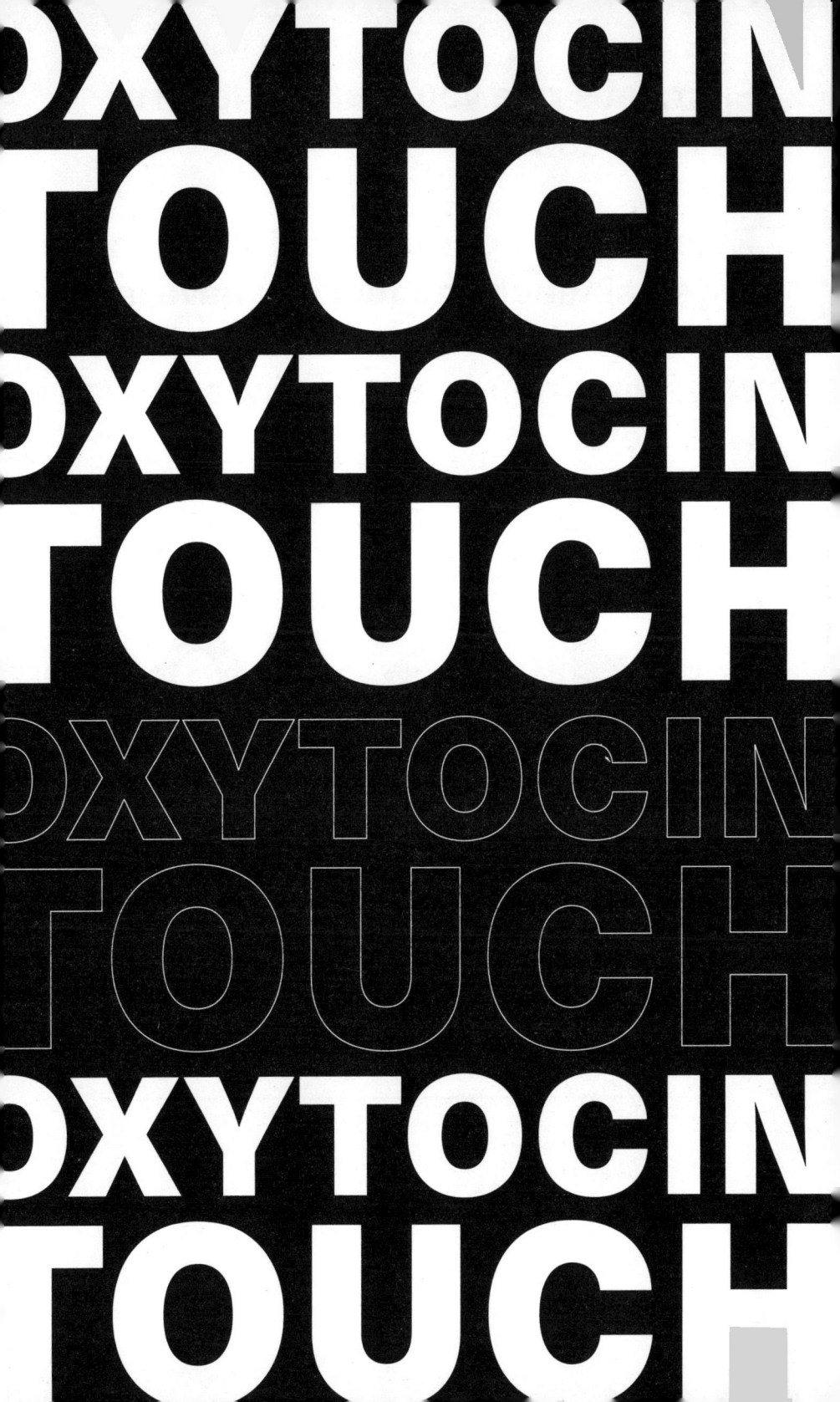

First, let's rate your current level of physical touch.

Rate on a scale from 1 to 10 how much physical touch you feel you have in your life. This includes in a romantic way, with friends and family, and even with pets.

1 ➔ 10

1 = none
10 = tons

Understanding TOUCH

Right at the very centre of the function of oxytocin is the deep desire it creates for physical safety.[26] During your first few months here on earth, you are welcomed with a huge amount of physical touch. Moments after you are born, you are placed on your mother's chest, and during these moments, oxytocin surges through you, contributing significantly to developing the emotional bond between you both.[27] Continued physical touch and other nurturing activities, such as breastfeeding (if it is an option), become integral components of your developmental journey.

Important note: *If you are unable to breastfeed, you can achieve this desired oxytocin release through intentionally increasing the quantity of physical touch you experience with your child.*[28]

As you navigate these first few months, you are only able to communicate through emotional expressions and sounds. In any moment of emotional distress, your primary caregivers come to you, they hold you, they cuddle, they rock you, and the safety of this physical connection calms you down. Fascinating modern neuroscience research has demonstrated that during these moments when someone is holding you, oxytocin is released into your bloodsteam and, as a result, your primary stress hormone, cortisol, reduces as well.[29] This is our first mention of cortisol, and it is an important component of our journey forward throughout *The DOSE Effect*. Many aspects of our modern lifestyles are leading to our cortisol levels (we can think of these partially as our 'stress levels') being incredibly high. We now know from neuroscientific insights that physical connection doesn't just build love and emotional connection, but it has a profound impact on calming your body down, too.

When you think about it, it makes a lot of sense. As you know very well now, we originally spent a great deal of time outside, in the wild, surviving. Moments of physical connection, whether it was in our early months of life, or at any point throughout our journey, would have provided a sense of reassurance and safety. The amount of **Touch** we are all receiving on a daily basis has been progressively reducing, particularly since the significant societal transformation we all went through as we navigated COVID. During the live DOSE training that I deliver in companies and schools across the world, we use live interactive questions that audiences can answer anonymously to support my understanding of them, and their understanding of one another. During our oxytocin session, I ask people the question, 'How many hugs do you average per day, from 0 to 10?' Before reading on, answer this question now: how many individual hugs do you have per day? This can be all with the same person, or all with different people; the total number of hugs is what we're looking for here.

When asking adults in companies this question, the average answer we see is between one and two per day. When asking young people in schools, the average answer we see is between none and one per day. This is simply not enough. The first time I ever asked this question live on stage, and saw the answer come out at an average of 1.4, it literally hurt my soul. I love humans, I know we need physical touch, and I was heartbroken seeing how little physical connection we all receive. At this moment I thought, 'Geez, these people need some hugs.' So, I asked everyone to stand up and go hug three people in the room. As I said this, I got a horrible look from the audience, a look that said, 'Godddd, why have you asked us to do that, that's so awkward?' I proceeded to guide them to do it. They stood up, they began hugging, and what I observed was crazy. People went from feeling awkward to the calmest and happiest I had ever seen them. When they sat down, they were laughing and smiling at one another. Watching oxytocin's impact live in action was phenomenal.

I have proceeded to do this exact exercise during every single oxytocin-training experience and the result is identical every time. The audience look at me, with their eyes saying, 'Whyyyy, I don't want to do that.' They stand, they hug, laughter and joy erupts through the audience, and the energy in the room shifts for the rest of the event.

When delivering this in schools, particularly with young girls, the impact I have seen when increasing Touch has been phenomenal. Many young people who message me via social media share with me their challenges around their friendships, bullying, and self-confidence. During this session, I set them a simple challenge, which we will come to in a moment, of aiming for a certain number of hugs each day with their friends. The feedback I've had from them about how this impacts how they feel in their friendship groups, their confidence, their reductions in anxiety, is honestly just beautiful to hear. We need physical touch; we need oxytocin.[30]

It is, of course, important to note that all of us have different preferences regarding Touch. Some people love a big hug, and others don't particularly enjoy it. If you are on the side of thinking that Touch isn't something you desire, it's important to get it in a way that feels comfortable for you. This might mean just occasionally hugging a particular person you feel safe around, or it could mean connecting with pets. People always put up their hands during this part of our live events and say, 'What about pets, what about pets?' You may have been pondering the same thing yourself throughout this last chapter. Rest assured: cuddling pets also provides an incredible oxytocin release.[31] So, whether Touch is something you love, or shy away from, connecting regularly with your cats or dogs is a great way to achieve the effect that we are seeking for here. And remember, earlier we mentioned that Touch leads to significant reductions in your stress hormones.[32] So, if you've had a particularly tough day, be sure to reach out and cuddle your partner, your friends, your kids, or your pets.

On the topic of cuddling, it is, of course, valuable to mention the relationship between oxytocin and your romantic relationships. Everyone who is reading this book will be in a different place. Some of you may be married, in relationships, dating, or currently enjoying a single period. Intimate physical touch is a moment that truly shows the power of this brain chemical. Take a moment to remember your first moment of intimacy with a partner you felt connected to. That first moment your hands touch and bodies touch, that electrical feeling that fires through your brain and body. That's oxytocin. That is a feeling we need.

FOR THOSE IN
Relationships

First, for those of you in relationships of some kind, I want you to really ponder how much physical connection you and your partner have. I am not referring here just to sex, although, of course, I hope sex is a part of your relationship. But I also mean holding hands, kissing, cuddling on the sofa, giving one another massages, or hugging when you say goodbye to one another. I do hear from lots of people that they are struggling with their sex lives and their sex drives. There are sections in many chapters in this book, including Gut Health, Deep Sleep, Exercise, Stretching, and Heat, which all have a positive impact on sex drive.[33] However, Touch is unbelievably important. I have a story from a recent training session that taps into this.

A man in a company I was training opened up to me when I was talking about Touch. He was in his mid-fifties and pretty averse to the fact that his company had put him on 'mental health training', which I totally understand. A big part of my life is sharing with people how much of a different approach DOSE is from what you typically might consider 'mental health training'. During my conversation with him, he shared his realization of how rarely he hugs his wife. I shared my thoughts and reassurance as to how often I am hearing this these days. I set him a simple challenge to ensure he hugged his wife properly every day before he left for work and upon his return home. This gentleman came back the next week with a huge smile on his face. I asked him how it went, and it was magical to listen to. He shared just how big an impact it had had on their connection, and not to get too personal, he also shared it helped reignite a part of their personal life that apparently hadn't been there for years. I'll leave you to figure that one out. What I'm saying is, please don't underestimate the power of something as simple as a hug.

We humans are a simple species with simple requirements that will allow us to function at our best. Whether you are wanting to create closer emotional connections or even improve your sex life, make sure you are deeply prioritizing the physical connection you are creating with that person.

FOR THOSE WHO ARE
Single

This is for those of you who don't have a partner. This is an important stage to go through and it is a great thing to experience. You need to take this as an opportunity to enjoy the experience of building a strong relationship with yourself and reach a point where you feel really comfortable and happy on your own. This will lead to you going into your next relationship in a much better position. Rather than entering a relationship to avoid loneliness, you are entering one because that person genuinely improves your experience of life. If you are in this position, physical touch is still important to give and receive. This means that when you see your friends, your family, your nieces, your nephews, anyone you have a connection with, you need to prioritize even more the importance of these hugs. Hold your hugs for longer, aim for three to five seconds, which provides a greater oxytocin release.[34] Making a conscious effort to create more physical connection with all the people you love will support you feeling far more fulfilled in this area.

Recent research has demonstrated the additional impact Touch has on people's perceived feeling of loneliness.[35] During our next chapter, Social Life, we will be discussing the importance of social connection, loneliness, and how you can feel more connected to the people in your life. Loneliness is something many people struggle with in our modern world. Touch is a key antidote to this challenge.

Making a conscious effort to create more physical connection with all the people you love will support you feeling far more fulfilled in this area.

Strategy

The focus of your strategy regarding Touch is to spend this week intentionally integrating more physical connection into your life. This will vary for different people. If you are in a relationship, this will include more kisses, more hugs, more cuddles, more massages, more hand-holding, and more sex. If you are not in a relationship, this will include more and longer hugs when you see your family and friends. It may also involve organizing for yourself to have a massage – this would, of course, be a treat, but a treat that would be incredibly valuable for your oxytocin and health.[36] Touch could also be a focus of your self-care routine; for example, moisturizing your body and applying cream to your face as part of a routine can also release oxytocin.[37] For anyone with pets, interact with them more this week, sit on the sofa and cuddle them, put them on your lap, or take them on more walks.

Challenge

I would now like you to complete the '**Touch Challenge**'. In order to complete this challenge, our goal is for you to have 5 hugs per day. I am very aware this may feel like a lot, but this is one of the DOSE challenges I have had SO MUCH positive feedback on and it wouldn't be a challenge otherwise!

You could hug one person five times. You could hug five different people, or you could do anything in between. This may involve a few hugs with your partner, a few with your family or friends, and a few with your colleagues. It may seem unusual, but, like everything in *The DOSE Effect*, I simply recommend you try it, and observe how you feel. Remember to hold them for a few seconds for the greatest oxytocin release.

Hold your hugs for longer – aim for three to five seconds, as this provides an even greater oxytocin release.

8

Prioritize Social Connection

First, let's rate your current social life.

Rate on a scale from 1 to 10 how much you feel you are socializing with others.

1 → 10

1 = never
10 = all the time

Understanding SOCIAL LIFE

The simplest way to know just how much humans crave and require social connection is to think back to the announcement of the first lockdowns in COVID. I want you to remind yourself of those absolutely ridiculous Zoom quizzes that took place. The horrific Wi-Fi. The desire to try and get everyone online at the same time. The challenges of everyone speaking at once. Despite the rather fast rise and fall of this Zoom quiz moment, it demonstrated one thing. Humans need each other. In that moment of lockdowns being initiated, we immediately devised strategies to remain connected. This is oxytocin.

During the last eighty-five years, the world's longest study on human happiness has been taking place. Beginning in 1938,[38] the Harvard Study of Adult Development[39] has tracked three generations of families to discover what truly contributes to our happiness. There are over seven hundred people in this study from a variety of economic backgrounds. Researchers have checked in with these people every year, asking them about their work lives, their home lives, and their health.

Dr Robert Waldinger is now the lead research scientist on this study, and the findings are fascinating. Contrary to how we may feel in society, wealth, fame, and success have not been the lead predictors of the contributors' health and happiness. In actual fact, the greatest predictor of long-term mental and physical thriving has been the quality of their relationships. People in the study who had closer relationships with their friends and family have lived significantly longer than those who did not. They have even exhibited greater cognitive functioning later in life, their memories staying sharper for longer than those who experienced loneliness!

Interestingly, it wasn't about how many social connections people had, or whether they were married all their lives, but instead it has all been about the quality of the actual connections. Those who rated themselves as the most satisfied with their relationships at age fifty were the healthiest at age eighty. Warm, loving relationships appear to buffer a range of the physical and psychological challenges that arise as we age. The findings from this study don't suggest that relationships need to be perfect. They may include arguments and challenges to navigate. However, the key predictive factor was that the individuals in the relationships deeply felt as if they could count on one another in times of need.

For thousands of years we lived in small, intimate tribes where social connection was at the forefront of our focus. As the modern world has developed, more and more of our time is spent behind screens, working and seeking pleasure. During this chapter, we are going to look at your **Social Life**, and how much of a priority it is for you. I, of course, recognize that all of us are uniquely different. Some of us may lean towards extroversion and love being in large social groups. Others may lean towards introversion and prefer more intimate one-on-one conversations. Regardless, one thing is clear: to thrive, we must prioritize building loving relationships in our lives.

Modern research exploring the role of oxytocin in social connection is fascinating. We know that early in life, the human connection we experience builds oxytocin.[40] This continues as we age. Moments of high-quality, focused social connection in both personal and romantic relationships are vital to stimulate this neurotransmitter in your brain and body.[41]

First, let's consider activities that enable good-quality social connection, then we will go on to how you can increase your oxytocin levels during these moments. As we now know from our understanding of dopamine, alcohol is one of the greatest challenges our modern world faces. A significant amount of social engagement involves alcohol. I know that if I drink less and have a more balanced relationship with my alcohol consumption, my life is better in every way. My relationships are stronger, I am far more productive, my mood is better, my energy is better, and my ability to maintain discipline in my life is far easier. While I appreciate that drinking alcohol is a part of our social culture, the activities I am going to suggest to you don't involve the consumption of alcohol and will provide you with alternative ideas that enable you to connect.

ACTIVITY 1:
EXERCISING TOGETHER

Consider your ancestors, and what they spent their time doing together. Building, walking, exploring, hunting, foraging. A great deal of our time here on earth was spent exercising together. As we will go on to discover in Chapter 16, Exercise, it is a vital ingredient for optimal physical and mental health. Exercise is a great way to build the strength of your social relationships.

Activities you may consider doing with a friend or family member include jogging, cycling, lifting weights, dance classes, yoga, Pilates, martial arts, and playing sport.

ACTIVITY 2:
WALKING AND RELAXING IN NATURE

As we will come to discover in my favourite and most transformational chapter in the book, Nature, walking in green spaces is a magical way to connect with someone. Nature has unbelievably calming properties for your brain and body, and being in these environments fosters great quality connection.[42] Interestingly a recent research study demonstrated that when people socialize in natural environments, they pay closer attention to one another and connect more deeply.[43]

Activities you could do in nature include walking, cycling, relaxing on the grass, or looking at and smelling different plants.

ACTIVITY 3:
LISTENING TO MUSIC TOGETHER

For thousands of years, humans have both created and listened to a wide range of musical sounds. During Chapter 18, Music, we will discover just how impactful music is on the health of our brains and bodies.[44] Listening to music as a social activity is a great way to connect. In comparison with watching TV together, music provides the space for conversation and connection. In a range of innovative studies, listening to music in social settings led to greater connection and enhanced mood.[45]

 Next time you are relaxing with friends or family, rather than putting on the TV, consider connecting through playing music instead.

ACTIVITY 4:
CALLING INSTEAD OF MESSAGING

A great deal of our social connection is now taking place via messaging one another on our phones. A fascinating study assessed the different levels of oxytocin released between messaging someone close to you or calling them. When simply messaging, there was no release of oxytocin at all. The comforting sounds of hearing voices provided a far more substantial release of oxytocin.[46] It may feel as though your need for social connection is being satisfied by messaging. Many of us feel this way. However, it is clear that we need to hear one another speak to bond in the very best way.

 When seeking to connect with someone you love, be sure to call or video call them.

ACTIVITY 5:
EATING AND DRINKING TOGETHER

Of course, one of society's favourite ways to connect socially is through our love for food and drink, whether this is having your friend or family member round to your home for dinner or meeting out and about for a walk and coffee. These moments are a brilliant way to connect. They are also where continuing your daily practice of Phone Fasting is essential. Sitting with someone in a coffee shop while they frantically reply to messages isn't the most enjoyable experience. In these moments, ensure that your phone isn't on the table and is on airplane mode. This enables you to focus on the conversation rather than acting as though the people on your phone are more important than the person in front of you.

Consider now which of the following social activities appear the most enticing to you:

1. EXERCISING TOGETHER
2. WALKING AND RELAXING IN NATURE TOGETHER
3. LISTENING TO MUSIC TOGETHER
4. CALLING INSTEAD OF MESSAGING
5. EATING AND DRINKING TOGETHER

TIPS
to strengthen your Social Connections

Throughout the next week, when you are in these social environments, I want you to consider these additional behaviours in order to truly optimize your oxytocin and strengthen the love and connection you feel.

1. REMOVE PHONES

Phones, specifically their notifications, provide a barrier to deep connection and conversation. Whenever you are with someone in a social setting, ensure your phone is not on the table and is away in a bag instead.

2. LISTEN ACTIVELY

This skill of paying close attention and listening when people are talking is essential and something so many of us struggle with. Have you ever found yourself in a conversation with someone where they are chatting away, but during this time, rather than listening, you are formulating your response to what they are saying in your mind? Then you are not truly listening to them, and are thereby failing to connect fully.

On the other side of this, when you are the one talking, it is so easy to know whether the other person is paying attention. You may have certain friends who you feel really listen to you; they make eye contact, they nod, and it feels as if they genuinely care about what you are saying. On the other hand, you may have some friends with whom you experience quite the opposite, when it always appears as if they are distracted or potentially seeking to move the conversation towards talking about themselves.

An incredibly simple way to strengthen the connection you experience in social moments is simply to actively listen when people are talking. It is a skill which, once you pay closer attention to it, will quickly improve. Your conversations will be better and more open. The person you are chatting with will feel safer, more comfortable, and more loved.[47]

3. COMPLIMENT

Have you ever received a compliment, be that for your parenting, your intelligence, your health, or your love for people? It is a magical feeling, receiving a genuine compliment. Modern neuroscience has demonstrated the benefits to our oxytocin system of sharing and receiving compliments.[48]

Compliments are something you can give more often. This is a wonderful way to deepen your connection with those you love. Compliments, of course, can often be about people's appearance. However, I think it is beautiful and important for us to also compliment people in different ways too. Here are a variety of ways in which you could compliment someone today.

- 'You have a beautiful smile that lights up the room.'

- 'You're so kind and empathetic to the people you love.'

- 'You're a great listener; talking to you always feels so relaxing.'

- 'Your positive attitude is contagious; it's always so much fun hanging out with you.'

- 'You have an amazing sense of style; you always look awesome.'

- 'Your determination and hard work inspire me massively.'

- 'You make the people around you feel so valued and appreciated.'

- 'Your intelligence and insight are unbelievably impressive.'

- 'You have a heart of gold; your generosity is magical to witness.'

- 'You have a wonderful sense of humour; your jokes always bring a smile to my face.'

4. MAKE EYE CONTACT

When in a conversation with someone, have you ever noticed the impact it has on the feeling of connection when you make direct eye contact with them? Whether this is during a conversation with a family member, a colleague, or in a romantic situation, you will feel the quality of your bond rise when you look one another in the eye. Interestingly, modern research has demonstrated that making eye contact during conversation is associated with higher levels of oxytocin.[49] In addition, individuals who make eye contact are perceived as significantly more confident, something many of us desire.

5. CONNECT PHYSICALLY

As we now know from the previous chapter, Touch, physical connection through hugging leads to significant rises in oxytocin and connection.[50] Ensure when greeting your friends or family that physical contact is a priority. This is also a good chance to take a longer hug if you haven't had much touch that day.

6. ASK GOOD QUESTIONS

Sometimes in social moments, it can be easy to sway towards predominantly talking about yourself. Ensure you ask the person you are with good questions, and show true interest in their life. This leads to someone feeling valued and significant – an important experience for all of us. A simple way to achieve this is through repeating things they have said, for example using names they have mentioned, and asking clear follow-up questions. This provides a clear sign that you have been actively listening.

SOCIAL CONNECTION
Checklist

1. Remove phone

2. Listen actively

3. Compliment

4. Make eye contact

5. Connect physically

6. Ask good questions

BUILDING YOUR
Social Confidence

Many people throughout my years of teaching have come to me having experienced social anxiety. This is a feeling of nervousness and self-consciousness in social situations. If you experience this, please consider following the guidance below.

During my time studying clinical psychology, I came across a concept called the '**Spotlight Effect**'.[51] This refers to the experience of being in social settings and overestimating how much other individuals are analysing your appearance or behaviour.[52] You may have experienced this yourself before, perhaps being in a social setting and wondering what people think of your outfit, what you're saying, or how you are coming across.

It is really important to note that this is very normal and natural behaviour for us as humans. It is usual for us to want to be liked and accepted by the group we are in. Being accepted and welcomed into a group has been a vital component of our survival for thousands of years. The challenge arises when we over-identify with these thoughts and find ourselves lost in them, and, as a result, disconnected from the conversation in front of us.

In these moments, our aim is to move our attention away from this internal voice in our minds and towards the people we are with.

Three steps to BUILD social confidence

STEP 1:
ACCEPT IT

The first step in navigating this is not fighting it, but rather telling yourself this is natural, and everyone experiences this. Fighting it and trying to stop it happening only exacerbates the problem. The funny thing about the spotlight effect is that people are often too busy wondering what you think of them, to take a second to analyse you. It's almost comical that in social situations we are all so curious and conscious of what others think of us, that none of us has any time to analyse one another. Accepting that you, alongside many others in the social setting, experience this, leads to you naturally calming down.

STEP 2:
MAKE EYE CONTACT AND STAND TALL

Making eye contact and standing with good posture has been shown to increase feelings of self-confidence and perceived confidence by others.[53]

STEP 3:
PAY ATTENTION AND CONTRIBUTE

This step is vital. As we know, a significant component of this challenge lies in the fact that we get lost in our thoughts. While accepting this state, making eye contact, and standing tall, it is imperative that you listen closely to the conversation in front of you and contribute to it. The more immersed you are in the conversation, the more your attention moves from internal analysis to external focus. This will naturally quieten the analytical thoughts in your mind.

Social Confidence Steps Summary

STEP 1 Accept it

STEP 2 Make eye contact and stand tall

STEP 3 Pay attention and contribute

In these moments, our aim is to move our attention away from this internal voice in our minds and towards the people we are with.

Strategy

Your focus over this upcoming week is to prioritize your Social Life and your relationships. Whatever headspace you are in right now, whether you are seeking to see a big group of friends or maybe you want to just call a family member or friend you haven't spoken to for a while, I want you to make social connection your priority. As we know, feelings of loneliness are an ever-increasing challenge in our modern lives. Ensuring you are creating the space in your schedule to connect with the people you love won't just boost your oxytocin levels but, as we know from the world's longest study on happiness, it will have a great, if not the greatest, possible impact on your wellbeing.[54]

Challenge

I would now like you to complete the '**Social Life Challenge**'. In order to complete this challenge, you must plan and experience three moments of high-quality social connection throughout the next seven days.

Ideas include:

- Exercising together
- Walking and relaxing in nature together
- Listening to music together
- Calling instead of messaging
- Eating and drinking together

And remember:

- Remove phone
- Listen actively
- Compliment
- Make eye contact
- Connect physically
- Ask good questions

9

Be Grateful Every Day

First, let's rate your current level of gratitude.

Rate yourself on a scale from 1 to 10 for how actively you feel you are considering what you are grateful for each day.

1 → 10

1 = never
10 = all the time

Understanding GRATITUDE

At the start of Part 2, I mentioned how oxytocin is not just created through the love and connection you build with the people around you. Oxytocin is also built through the love and connection you build with yourself.[55] Throughout the next two chapters, Gratitude and Achievements, we are going to explore the relationship you have with yourself. This refers to the conversation taking place in your mind each day.

Many of us struggle with an inner voice that is critical, that identifies your personal flaws alongside the difficulties taking place in your life. It is important to understand that this is a natural and evolutionarily advantageous characteristic. Having a mind that can identify where it is making mistakes and work to make progress is valuable. However, this mechanism, in a world of social media – where we are constantly being shown alternative ways to look, feel, and live – provides an overwhelming level of analysis of ourselves and our lives.

'Comparison is the thief of joy.'

This is an iconic and accurate statement.[56] Many of us spend a great deal of time thinking about what we don't have, whether that's wealth, homes, holidays, clothes, cars, looks, or experiences. With comparison, simply put, being the process of thinking about what you don't have, the only viable solution to this is **Gratitude**, which makes you think actively about what you do have.

I know that society is very aware that 'we need to be grateful'; it's a topic that has been discussed at great length. But ask yourself at this moment, do you have a clear, daily practice where you deeply consider what you feel fortunate to have and experience? Is the answer yes? In my opinion, gratitude is a non-negotiable strategy that is required to mentally thrive in our world today.

Let's consider our ancestors once more. Imagine you had one tribe that was flourishing; they had a great method for finding food, collecting water, building shelter, and ultimately surviving. They felt peaceful and happy with their experience of life. Say I then turned up to this tribe with an iPad and shared with them a video of three other tribes who had a more proficient method for finding food, collecting water, building shelter, and surviving. Very quickly the first group would enter a state of envy and disappointment because of how their life compared with the video. In our modern world, we are living in this state, constantly being shown a better way to experience life.

Let's return to the equation we discussed at the beginning of Part 2:

HAPPINESS
is equal to
reality
minus
expectations

If our expectations for our lives are too high, and above the reality we are experiencing, we will live a life of disappointment and envy. Gratitude is the solution to this challenge, one that is easy to implement and monumentally impactful on the health of your mind.[57] Gratitude is a practice that is of huge value in two aspects of your life: your relationships and experiences with others; and the conversation that takes place in your mind every day about your life.

These TWO AREAS are sharing gratitude and feeling grateful.

1. SHARING GRATITUDE

A recent study found that gratitude practices in relationships, where partners intentionally exhibited greater levels of gratitude towards one another, were associated with greater levels of oxytocin. The researchers stated that oxytocin and gratitude are 'the glue that binds adults into meaningful and important relationships'.[58]

If today, your friend, partner, colleague, or child came up to you and simply said 'thank you' – 'Thank you for how you support me in my life, for how you show up for me, for cooking amazing meals, for working so hard, for cleaning our home, for listening to me', or whatever it may be that aligns to how you contribute to their life – you would likely experience a beautiful, warm feeling inside. A feeling that you are seen and valued for all you do for that person.

A challenging feeling can arise within us when we feel as though we are putting a great deal of effort into our relationships and that effort is not acknowledged or seen. I hear this all the time, specifically when I am training companies, where people feel as if they are putting a huge amount of effort into their team and driving towards the company's goal, and that effort is not necessarily respected or seen. I observe the same when training families, where individuals' contributions are not acknowledged and recognized. A particular time I see people experiencing this is during motherhood, when mothers' unbelievable contributions are not necessarily given the correct level of acknowledgement.

It is vital we develop the ability to express these emotions of gratitude on a regular basis. Today, for example, when you see your family, your friends, your colleagues, simply say thank you. Thank them for how they are impacting your life. Make them feel seen and noticed. During this moment, both of you will experience an incredible rise in oxytocin and your bond will deepen.[59]

2. FEELING GRATEFUL

A term known as 'dispositional gratitude' refers to the ability one has to notice and appreciate the positive aspects of one's life. This ability is associated with significant improvements in one's wellbeing alongside greater levels of oxytocin.[60] One of the great difficulties we face in modern times is our awareness of all the challenges that face our world each day. Every day we hear news of terrible events taking place. It is incredibly important to understand that the human brain is a learning machine. It gets conditioned incredibly easily. When we are seeing and hearing that the wider world is going wrong, via the news, it is very easy for this to affect how you begin to look at your own life too. You may find at times that your thoughts orient towards the negative, towards what isn't going to plan in your life. This is where 'dispositional gratitude', the art of noticing and appreciating the positive experiences you have, is absolutely fundamental.

I want you to take a moment now to consider what you are grateful for in your life. Very quickly you may think: my family, my home, my friends. Our aim here is to take this to another level, and think as specifically as we can about each area. An example of this may be, instead of thinking, 'I'm grateful for my family', ask yourself, 'Who in my family do I feel grateful for at this moment?' Okay, now you have a specific person in mind. Ask yourself, 'Why do you feel grateful for them?' Is it something they have done to support you recently? Is it their positive energy? Is it simply the joy of getting to connect with them? The more specific you can be, the more immersed your mind becomes in the state of Gratitude we are seeking to experience.

Here is a range of different aspects of the human experience that you could feel grateful for. Take a few minutes to read through them, and select three that you deeply connect with at this moment.

I'm someone who goes through challenging days and low moments. These might arise as a result of relationship challenges, work challenges, exhaustion, over-engagement with 'quick dopamine', or sometimes my emotions naturally fluctuate. Whatever the cause may be, I have found Gratitude to be one of the most profound solutions to alter my state of mind and bring me back to a state of energized peacefulness.

I am GRATEFUL for...

1. A specific friend or family member
2. My home and living environment
3. My health, be that my ability to move, or feelings of energy and vitality
4. My financial stability
5. My opportunities and new experiences that are becoming possible for me
6. Nature, and the beautiful world we live in
7. My nutrition, what I have the opportunity to eat and drink each day
8. My learning, the opportunity to expand my understanding, with this book, for example ☺

Strategy

Gratitude is a practice that will improve how you feel the moment you put it into action. However, our focus here is to make this an integral component of how you think and act every single day. This is certainly a skill that applies to the compound growth rule within DOSE – where all the small changes you are making to your life add up and multiply to create true psychological transformation. The more consistently you do it, the more impactful it will become over time.

There are two points in your day when I would guide you to intentionally check in with yourself and have a moment of Gratitude.

1. MORNINGS

Gratitude at the start of your day is an effective method to make you more optimistic.[61] When integrating a new habit into your life, it is important to pair it with something that is already a key part of your daily routine. You could, for example, do this when you first wake up, as you're making your bed (Discipline), when you are in the shower, or when you are brushing your teeth. My greatest advice to you would be to do this on a morning walk, prior to seeing any social media. During Part 3, Serotonin, we are going to discover the transformational impact of sunlight and nature. Pairing your Gratitude practice with a morning walk is a magical way to start your day and will help you to begin in the very best frame of mind.

2. BEDTIME

Lying in bed at night can sometimes be a moment when our thoughts switch on. You may have found yourself wanting to have a TV show or podcast on as you begin trying to fall asleep, to settle a noisy mind. We are so unbelievably stimulated all day, and going from that level of stimulation to a state where we are in quiet rest can be challenging. Many people share with me that, as they lie in bed at night, worries can arise in their mind about aspects of their lives that aren't going to plan. This is your second moment in the day to activate your Gratitude practice. In this moment, moving your mind towards this state is going to calm it and your thoughts as your brain will be reassured that you are safe and that everything is okay. Potentially, you might actually feel that you are better than just okay; there may be wonderful things occurring in your life that your mind had forgotten in that moment. Whether you are someone who experiences a noisy mind at night, or you feel peaceful and want to build upon that state of mind, Gratitude as you fall asleep is essential.

Remember, alongside this intentional practice, be grateful in moments when someone supports you in your life, your work, with your home, your family, your relationships, whatever it may be. Share your Gratitude. Ensure that the individual feels recognized for their love and care for you.

Your daily gratitude question

This is the simple part. To successfully practise Gratitude, simply ask yourself:

'In this moment right now, what is the number one thing I feel most grateful for?'

Examples

- 'I feel really grateful to have . . . in my life'
- 'I feel really grateful for the home I live in'
- 'I feel really grateful that my body is healthy'
- 'I feel really grateful for the opportunities I have in my life'
- 'I feel really grateful for the beautiful nature in our world'
- 'I feel really grateful for the financial situation I am in'

Upon asking yourself this question, consider the variety of options we explored on page 149, and select the one that is most prominent in your mind. Then spend a few minutes asking yourself, 'Why do I feel grateful for this?', 'How is this impacting my life?', 'What would it be like if this person didn't exist, or if I didn't get to have this thing or experience?' Immerse yourself in this deep state of Gratitude twice a day and you will experience a seismic shift in the positivity of your thinking each day.

Challenge

I would now like you to complete the **'Gratitude Challenge'**. In order to complete this challenge, you must ask yourself the 'Daily Gratitude Question', once every morning, and once every night for the next seven days.

As you complete this challenge, ensure your gratitude towards others is also at the forefront of your mind.

10

Believe in Yourself

OXYTOCIN
ACHIEVEMENTS
OXYTOCIN
ACHIEVEMENTS
OXYTOCIN
ACHIEVEMENTS
OXYTOCIN
ACHIEVEMENTS
OXYTOCIN
ACHIEVEMENTS
OXYTOCIN

First, let's rate your current level of self-belief.

Rate yourself on a scale from 1 to 10 for how much you believe in yourself.

1 ➔ 10

**1 = not at all
10 = huge self-belief**

Understanding ACHIEVEMENTS

I want to start this chapter by telling you a couple of stories. Stories that provide insight into the importance of your inner voice.

I want you to imagine two sets of parents. Both sets of parents are currently raising a young child, let's say four or five years old. The first parents choose to raise their child by constantly identifying and telling them where they are making mistakes in their life. 'You're doing this wrong in school, this wrong with your friends, this wrong at home.'

The second parents decide to make a conscious effort to identify where their child is getting things right. 'This aspect of school is going great for you, it's so good that you act in this way with your friends, thank you for how you help in this way at home.' Think about the progress and self-belief of these two children. The first one would be incredibly unconfident and unsure of how they should behave or what they are skilled at. The second one would believe in themselves and the skills they possess. This is simply as a result of the style of communication they experienced from their parents, nothing else.

When we look at how we talk to ourselves in our mind, often it leans towards the style of those first parents, constantly criticizing ourselves for where we are falling short in our lives and making mistakes. We rarely identify and celebrate our successes. In this chapter, we are going to develop a simple, daily practice that will alter your style of communication with yourself and significantly increase how much you believe in yourself.

Our second story doesn't just look at how positive self-talk is important from a self-belief point of view, but also the likelihood of achieving the goals you are seeking to attain. Imagine you have two people with the intention of wanting to eat healthier and treat their body better. It is a Monday morning and they both decide they are going to have a really healthy week. Monday to Friday, they both do amazingly, fuelling their body with valuable, healthy nutrients. On Saturday, however, both of them let go of their healthy week and eat a giant pizza and tub of ice cream.

One of these people gets angry, telling themselves how they 'always do this', how they're 'terrible at dieting', and how they are 'never going to be able to be someone who eats healthily'. The other person takes a different route. They recognize that it's frustrating that they let go of their healthy week. However, they take a moment to consider how incredible it is that they just had five consistent days of eating healthily. They celebrate the achievement of doing so.

The first individual has reinforced the negative behaviour. Their brain has heard, 'I'm terrible at eating healthily,' over and over again. The second person has reinforced the positive behaviour: 'It's incredible that I had five days of eating healthily.' Despite both of them doing the exact same thing from a nutrition standpoint, over time they are going to experience radically different outcomes. In order to successfully integrate a positive habit into your life, you must also develop the ability to notice and regularly celebrate your progress towards it.

Fascinating neuroscientific studies have demonstrated this relationship between oxytocin and a person's ability to communicate with themselves in a positive and constructive way.[62] One recent study even found that higher levels of oxytocin led to a reduction in negative inner communication in men who were navigating anxiety.[63] In addition, identifying and celebrating one's achievements go beyond merely how you communicate with yourself; they also influence how you communicate with others. When observing the impact of oxytocin in group settings, oxytocin has been revealed to be associated with rises in people's ability to share positive emotions, trust one another, cooperate better, and ultimately create greater group cohesion.[64]

Before we develop your strategy to identify your **Achievements** and those of others, it is important that you understand the basis of neuroplasticity. Neuroplasticity is defined as 'the ability of the brain to form and reorganize synaptic connections, especially in response to learning or experience'.[65] A simple example of neuroplasticity in action would be brushing your teeth. This is a skill that is deeply patterned into the neural wiring of your brain. If tomorrow I said to you, for the next month I need you to brush your teeth with the other hand, throughout the first week this would be very challenging. However, as the month went by, you would develop the capacity to brush your teeth with the other hand. This is neuroplasticity in action, your brain rewiring itself. If I told you to continue with this new hand for three months, you would find that after that period of time it would actually be a challenge if you went back to your original hand.

When you consider how you speak to yourself, it's likely you have a great deal of negative and critical thoughts about your work, your appearance, your relationships, your success, whatever it may be. The only solution to this is creating a new voice in your mind. One that celebrates you. Eventually, once you have done this for long enough, you will start to find it hard to criticize yourself.

Strategy

To integrate this skill into your life, in a similar way to Gratitude on page 149, we need to select a moment in your day where you will identify your primary recent achievement.

1. MORNINGS

My guidance would be to pair this activity with your Gratitude practice. First, ask yourself your Gratitude question: 'In this moment right now, what is the number one thing I feel most grateful for?'

Then ask yourself your Achievements question (see next page). You can ask these questions in the morning while you are making your bed, showering, or brushing your teeth, but I would highly recommend that they are asked during your morning walk (see page 94). I am someone who found walking in the quiet incredibly challenging and it is something I very intentionally avoided at all costs. Both Gratitude and Achievements became the solution to this. During our next chapter, Nature, we will discover the true power of immersing yourself in natural environments, headphone free.

2. AFTER WORK

An additional moment when you can take a minute to identify your primary achievement is when you finish work. For many of us, our working lives are intense, fuelled by a never-ending list of tasks that need to be completed. It can be easy to spend a day working incredibly hard, and then still find the voice in our minds telling ourselves all about what we didn't manage to complete. At this moment, as you are leaving work, take a few minutes to celebrate yourself for the progress you made on the projects you are working on.

Important note: There is a very significant additional benefit to the identification and celebration of your Achievements on another brain chemical, dopamine! During Part 1, we discussed the importance of My Pursuit, the selection and pursuit of the primary goal in your life. At moments when we make progress towards this goal, and notice such progress, we don't just build our oxytocin, but we also build our dopamine as well[66] – increasing our motivation levels to continue striving to attain the goals we are seeking.

Your daily achievements question

This part is simple. In order to successfully identify your Achievements, ask yourself:

'In this moment right now, what is the number one achievement I feel most proud of?'

Upon asking yourself this question, consider what you have achieved that made you feel proud. Remember, this doesn't need to be anything huge. These are just daily achievements you feel are supporting you and ensuring your life is heading in a positive direction.

EXAMPLES OF DAILY ACHIEVEMENTS INCLUDE:

1. Spending time away from your phone
2. Focusing better when you are working
3. Spending time reading this book
4. Connecting with your family and friends more
5. Doing something kind for somebody else
6. Having more physical touch in your life
7. Being more grateful every day
8. Talking about your appearance in a more loving way
9. Spending time in natural environments
10. Exercising more regularly
11. Consuming healthier foods and drinks
12. Maintaining a more organized home

Once you have identified your achievement, celebrate yourself for it. This may seem unusual at first, but just like a loving parent, tell yourself, 'Well done', 'It's awesome you've been working on this,' 'The benefits to my life have been . . .' and so on. Consistently providing yourself with this positive feedback about your daily behaviour leads to a huge alteration in the positivity of your self-talk, and ultimately it leads to far greater levels of self-belief. Rather than constantly telling yourself where you fall short in your life, you start to realize, wow, I am making progress. And progress breeds more progress. You gain momentum and your life, your health, and your habits improve quickly.

Challenge

I would now like you to complete the **'Achievements Challenge'**. In order to complete this challenge, you must ask yourself the 'Daily Achievements Question' once every morning and once after work, every day for the next seven days.

As you complete this challenge, ensure your capacity to identify and celebrate other people's achievements is also at the forefront of your mind.

IMPORTANT NOTE: *When you notice that you experience negative self-talk, come back to your achievements and remind yourself of the progress you have made recently. Think of it like a see-saw: if negative self-talk arises, you need to send the see-saw up the other way, with positive self-talk.*

Building Your DOSE

Throughout Part 2, we have explored the true power of building your oxytocin levels to create more love in your life.

You have been on an awesome adventure of experimenting with a range of new actions, which will support optimizing this system in your brain.

I now want you to take some time to consider which one of the five primary oxytocin actions feels the most important for you to continue to prioritize. I, of course, believe it would be incredible if all of these behaviours remain a priority in your life. However, selecting one core behaviour and ensuring it becomes deeply embedded in your life is essential. Add them to your routine one by one if that is most helpful to you.

WHAT WILL BE YOUR PRIMARY
Oxytocin Action?

1. CONTRIBUTION
Ensuring that serving others is at the forefront of your mind

2. TOUCH
Prioritizing increasing the amount of physical connection you experience

3. SOCIAL LIFE
Making time to connect deeply with those you love

4. GRATITUDE
A daily practice to immerse your mind in the joy of your experiences

5. ACHIEVEMENTS
A commitment to celebrate yourself for the effort and progress you are making in your life

Be sure to chat with a friend or family member about the primary oxytocin challenge you have selected!

PART 3

Feel Energised and Happier

Understanding SEROTONIN

Welcome to Part 3 of your DOSE journey, Serotonin. Wow, I'm excited for this one – this is the part I have been buzzing to write. Serotonin is a magical chemical that, when activated correctly, empowers you towards a much healthier experience of life. The best way to understand serotonin is to think of it as the chemical that will help you to take good care of your body.

In our modern society, we have created the idea of 'mental health'. This can often lead to us feeling as though the challenges that occur in our thoughts are only happening in our minds. However, this isn't the reality. Instead, having a calm, healthy, energized body is essential to experiencing true joy within our minds.

Throughout Part 3, we will be exploring the true function of serotonin and how many aspects of our modern lives are leading to a reduction in its activation, before looking at a range of activities and challenges for you to explore that will enable you to optimize your serotonin.

The Serotonin Principles

PRINCIPLE 1:
90 PER CENT IS CREATED IN YOUR GUT

The most important aspect of serotonin for you to understand is that this chemical is not entirely created within your brain. In fact, 90 per cent of it is produced within your gut![1] This is incredibly different to the three other chemicals we are learning about, which are predominantly produced within your brain.

PRINCIPLE 2:
THE HAPPIER YOUR BODY,
THE HAPPIER YOUR MIND

The serotonin that is produced in the gut has been shown to directly impact our mood, energy, emotions, and nervous system function.[2] Given that it is produced within your gut, this provides a very clear insight into the importance of caring for this part of your body. The happier your body, the happier you will feel in your mind.

Your emotions are messages from your body

The word 'emotion' literally means energy in motion. For many of us, when we experience challenging emotions such as sadness, worry, or low mood, we feel uncomfortable and we will try to distract ourselves from these feelings, believing it is the easiest route to take. An example would be feeling sad and choosing to eat some sugary food, or to scroll through social media to take your mind off the emotion. As we know from Part 1, Dopamine, engaging with these 'quick dopamine' behaviours only exacerbates the problem. As we work through this part on serotonin, I want you to keep an important concept in mind: these feelings are coming up within you for a reason. When you observe them closely, you may notice a lot of them are 'gut feelings'. They feel like they are coming from your body, and they are!

They are simply messages that are trying to shift how you are living your life. Your brain and body are incredible at survival: that is their primary objective, to stay alive and pass on your genes. Just as healthy behaviours provide a rewarding feeling within you to reinforce doing them more regularly, unhealthy behaviours provide a negative feeling, to try and stop you engaging with them so often. Start listening to your emotions. In Chapter 11, Nature, and Chapter 14, Underthinking, this concept will become clear. The more you listen to and align your daily behaviour to your emotions, the happier and healthier you will become.

Understanding the brain–body connection

Each of us have twelve cranial nerves these are the primary nerves that start at the top of your spine and communicate with your brain to cover various functions such as eyesight, taste, hearing, and much more.

Eleven of these nerves go from the top of your spine and travel upwards into your brain. One nerve travels downwards. It makes its way down into your throat, then into your chest, then into your abdomen. This nerve is called the vagus nerve. It was given this name from the Latin word 'vagus', which means 'wandering', because of its extensive connections around your body. The vagus nerve enables communication between your gut and your brain, and is constantly assessing the state of your body. For example, it monitors your heart rate, breathing, digestion, mood, energy levels, and immune system.[3] If your body is being treated in the healthy way that it desires, this will have a direct impact on how you feel emotionally. This is where things get really interesting, because the term 'mental health' leads us to think this is all something occurring in our brains. It is clear, however, that your body also has a huge impact on the experience you have in your mind each day.

The Vagus Nerve

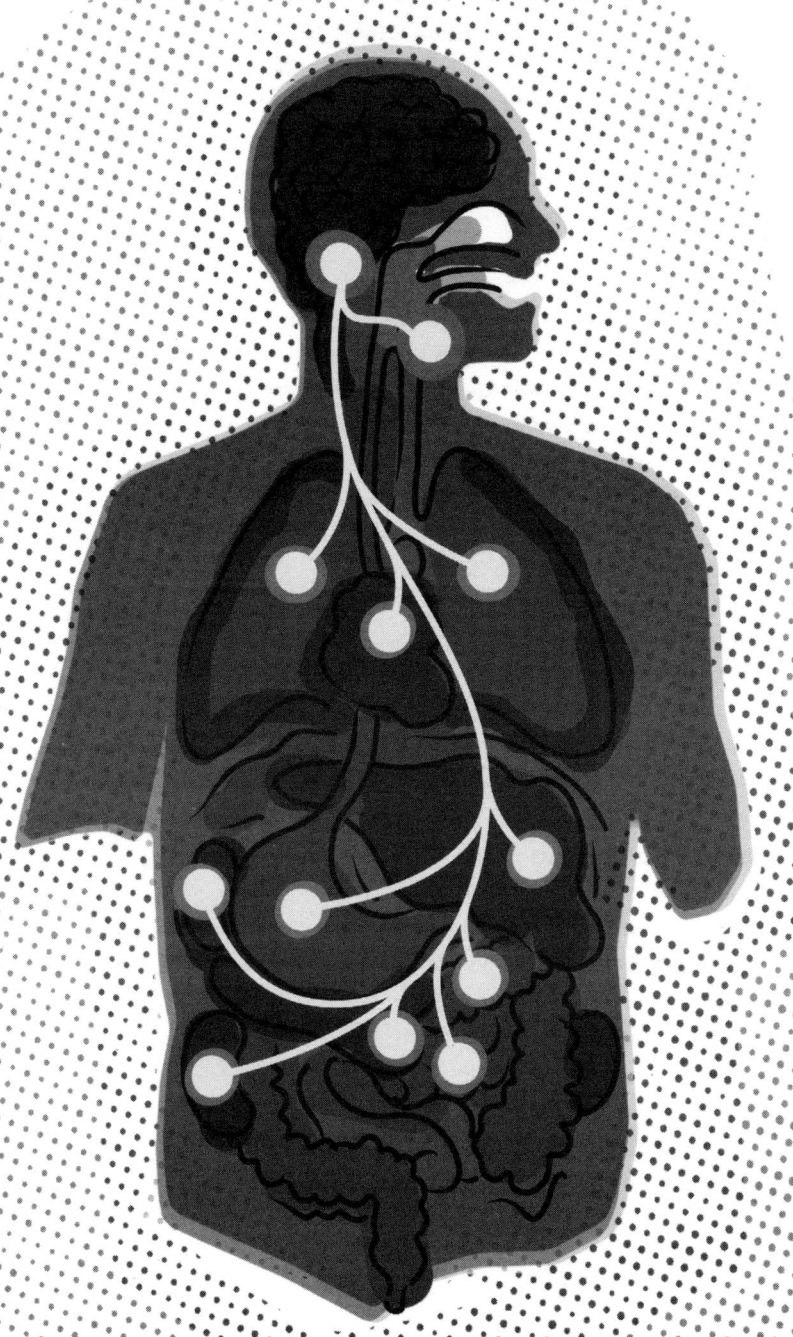

There are two primary functions I want you to learn to associate with serotonin. These are your mood and your energy levels.[4] It is important to recognize how connected these two feelings are for us all. The more you pay attention to your mood and your energy, the more you will see their connection. If your energy levels are very low, it is very hard to maintain a good, calm, positive mood. The problem is, our modern world and lifestyles can be incredibly tiring. Therefore, throughout this chapter we are going to find ways to optimize your energy levels and, as a result, lift your mood.

In our ancestors, serotonin would have been optimized to a high level. We know their dopamine was fulfilled through the constant pursuit of survival.[5] Their oxytocin was fulfilled by their deep requirement for love and connection as groups.[6] With their serotonin, we must consider how their bodies were treated. I want you to really visualize their lifestyle, which was sometimes hard but essentially very different to ours. They would wake up in the morning to bright, natural light, while being immersed within a natural environment. They would hear the birds, the animals, and the sound of nature around them. They would explore their local environment, in the pursuit of food, water, tools, shelter, or new places to live. In order to survive, they built a connected relationship with nature. They would eat unprocessed food from the land, and drink fresh water that came from the rivers and mountains around them. They would sleep deeply at night with no access to artificial light. This is what serotonin wants. When we compare their daily lifestyle with our way of living now, you can see how there are some key causes of low serotonin today. Let's explore how low serotonin would feel and then what causes this to occur.

Do you have low serotonin levels?

If your serotonin levels are low, you will often experience nervousness, anxiousness, low mood, or low energy levels.[7] Remember, it is not unusual if you are struggling with these symptoms; it is natural and many of us are. The key is gaining awareness and then beginning to incorporate behaviours that provide the solution. Now consider which of the following four causes of low serotonin may be impacting you.

The FOUR CAUSES of low serotonin

1. **An unhealthy processed diet.** This one is simple to understand. Serotonin is created in your gut.[8] If healthy, nutritious foods arrive in your gut, your gut thinks 'great, I can easily create serotonin from this food'. However, if unhealthy, sugary, fatty foods arrive in your gut, it has to spend time simply trying to eradicate those foods from the body. Building serotonin becomes the last of its priorities. I'm sure you've experienced this yourself. You eat some unhealthy food, you experience a huge spike of dopamine in your brain from the sugar and initially it feels amazing.[9] After you have finished eating, as your body digests the food, you experience a dip in your mood and energy levels[10] and you crave more processed foods for another hit. Throughout Chapter 13, Gut Health, we will explore your entire relationship with food and drink and how we can improve the health of your gut.

2. **Lack of good-quality sleep.**[11] This is now a common problem. Many of us have highly demanding jobs and, alongside that, deep addictions to our technology (I include myself here). Staying up late in the evenings on our phones, checking them at night, and scrolling through them as soon as we wake up creates a huge range of challenges for your neurobiology. In Chapter 15, Deep Sleep, you will discover the precise steps you can take to improve how quickly you fall asleep, how long you sleep for, and the quality of the sleep itself.

3 Lack of time in nature.[12] Unsurprisingly, having evolved over 300,000 years traversing the natural world, your neurobiology is deeply designed to experience a close connection with the natural world and a great deal of sunlight. Our digital way of living is progressively leading to the majority of our time being spent indoors. Alongside this, more and more of us are not prioritizing spending time in natural environments. In Chapters 11 and 12, we will begin your journey to a new relationship with the natural world around you.

4 Lack of sunlight. This, of course, is caused by lack of time in nature. In the Sunlight chapter, we will learn that any exposure to daylight, whether you're in nature or not, is incredibly beneficial for your serotonin levels.

How high serotonin levels will feel

Once you begin incorporating your chosen serotonin-boosting activities, the kind of symptoms you will experience are being in a great mood, feeling calm, and feeling far more energized every day.[13] I am excited for you to experience this.

On the next page you will find a simple illustration that provides the key functions, principles, feelings, and behaviours that are associated with optimizing your serotonin.

Serotonin Summary

Function →
- Mood
- Energy

Principles →
- 90 to 95 per cent is created in your gut
- The happier your body, the happier your mind

Low Serotonin Symptoms →
- Anxious
- Tired

Low Serotonin Causes →
- Unhealthy food
- Lack of sleep
- Lack of nature
- Lack of sunlight

High Serotonin Symptoms →
- Good mood
- Energetic

Serotonin Actions →
- Nature
- Sunlight
- Gut health
- Deep sleep
- Underthinking

11

Rediscover the Natural World

SEROTONIN
NATURE
SEROTONIN
NATURE
SEROTONIN
NATURE
SEROTONIN
NATURE
SEROTONIN

First, let's rate your current connection to nature.

Rate on a scale from 1 to 10 how connected you feel to the natural world.

1 ➔ 10

1 = No connection
10 = Maximum connection

Understanding NATURE

Nature has been the most significant factor for me in achieving strength both mentally and physically, and it has definitely helped me to thrive.

A big statement I know, but one made for a significant reason. For context, as a young kid I enjoyed going to parks and playgrounds, playing in forests, and swimming in the sea. However, as I grew up and entered my teenage years, spending time in **Nature** became very low on my list of priorities.

It wasn't until the onset of COVID that everything changed. During this period, I went to stay with my then girlfriend and her parents. They lived in a more rural, natural setting. During that time, I was really struggling with my addiction to my phone. I was waking up and scrolling, checking my phone all the time, even when I was trying to be productive with my work. I was even scrolling in the evening when I should have been connecting with the people I was with.

I decided to set myself a challenge of going for a sixty-minute walk every morning when I woke up, around the local fields, without my phone, and, very importantly, without checking my phone before I left. To me this was an unbelievably big challenge. It was a difficult prospect. This was because I had a fear of having to listen to all the thoughts in my mind with no distractions as well as a fear of boredom and lack of stimulation. I thought through my average day. I would wake up, and put music or a TV show on to play in the background. If I ever walked anywhere, I would have my headphones in with music or a podcast on. If I was relaxing in the evening I would be watching TV, and if I was trying to fall asleep I would have something playing in the background. My mind didn't enjoy the idea of listening to just my thoughts for any period of time and found living a life of distraction far easier.

By setting myself the challenge of a no-distraction walk, I quickly discovered that being in nature, in the quiet, was truly life-changing. It is one of the main reasons why I have written this book and I am running my business. I now feel happy and calm with how my life is going.

MY STORY

I am going to bring you along to experience one of these walks and show you how integrating this into your routine could transform your experience of life too.

I wake up; it's a cold, bright morning. I work hard to resist checking my phone. I head to the bathroom, I brush my teeth, splash some cold water on my face, and then get ready to set off for my walk. I step outside and begin walking. For a few minutes I think, wow, this is lovely, nature is pretty cool. Quickly, however, other thoughts arise in my mind, thoughts about some of the challenges in my life, whether they be in my work, my relationships, my health, my addictions, my level of success: anything negative. As I spend more time in the silence, more of these thoughts arise, often highlighting where my life isn't necessarily going to plan. I begin to really ponder on this walk how I never spend any time in the quiet. I really begin to consider how I must develop the ability to spend quiet time alone in order to create a good, happy life. I have to be able to be good friends with myself.

I then began to develop strategies to occupy my mind when in the quiet and when my mind becomes filled with negative thoughts. I was kinder to myself and created a more positive state of mind. When I began looking at the research behind spending time in nature, I saw the phenomenal impact that occurs when we become truly present in natural environments. It became clear that spending time in these natural environments would boost my serotonin levels.[14]

Therefore, I knew that if I could spend more time there, particularly in the mornings, I would be in a better mood and have more energy for my day ahead.[15]

I have since discovered that the benefits of nature go beyond boosting your serotonin levels. I came across an incredible framework known as the Attention Restoration Theory or ART.[16] The ART discovered that spending time in natural environments not only boosted your wellbeing, but it also found our ability to concentrate can be restored by nature. As we know from our discoveries in Part 1, Dopamine, many of us struggle with our capacity to concentrate. Over thirty-one separate research studies have been conducted on a range of different groups of people and all of them revealed that consistently walking in natural environments restores our capacity to concentrate.[17] Our digitally based lives are incredibly demanding on our attention, constantly having to engage a lot of cognitive effort. Mental fatigue is something that so many of us experience. The restorative benefits of spending time in nature for your brain and body cannot be overestimated.

With a clear understanding that spending time in nature would be transformative for my life, I then had the challenging task of figuring out how I could incorporate this into each day. Not only did I want to work on my phone addiction and my ability to spend time in the quiet, but I knew that the only way I would truly experience these restorative benefits was if my full attention was immersed in the environment around me and not distracted by music or a podcast.

As I walked out each morning, pondering different thoughts, I began considering how I could incorporate different aspects of DOSE into my experience. As I mentioned, in these quiet spaces, it can be easy for the mind to wander and think about your challenges, or what isn't going to plan in your life. This is where the skills of Gratitude and Achievements really began to impact my life for the first time, as I would focus instead on feelings of gratitude and my achievements while I was walking. What I very quickly began to notice was that the walk was far more enjoyable, and day by day, my mindset became more positive and more optimistic about my life, too. Once in a more positive headspace, I began to truly ponder My Pursuit for the first time, carefully thinking through what I wanted my future to look like. Doing this each day is what I believe made it become a reality.

While spending time outside, I also wanted to see if I could develop more of a connection to nature. I had heard people say how much they 'loved nature' and how attuned to it they were. If I am honest, this had always confused me. I began exploring, focusing on each of my senses. First, I tried really paying attention to what I was looking at, intentionally observing all the different colours, plants, trees, and views around me. Then I would tune in to the sounds I could hear. After this, scent. I would take deep breaths and notice the natural scents around me.

> **Nature Checklist**
>
> **1. SIGHT**
> Count how many different colours you can see
>
> **2. SOUND**
> Listen closely to the different sounds you can hear all around you
>
> **3. SMELL**
> Breathe in deeply as you walk and think about what you smell

I found that when I walked among pine trees in particular, the smell would immediately make me feel good. I went home and researched this and discovered that all plants emit natural chemicals called 'phytoncides'.[18] These have antibacterial and antifungal properties to help protect the plants from disease. When analysing what occurs in our bodies when these phytoncides are breathed in, researchers found that a very significant rise in something called 'natural killer cells' occurs too.[19] Your natural killer cells play an integral role in your immune system.[20] Upon looking into this, I found that coniferous trees emit a particularly large amount of these phytoncides,[21] which is why they were making me feel particularly good.

I came across a fascinating Japanese concept called 'Shinrin-Yoku',[22] which translates to 'Forest Bathing'. In the early 2000s, the Japanese noted a significant alteration in people's mental health and coined a term called 'karoshi',[23] which referred to an individual who was working extreme hours and spending too much time in city environments. Upon exploring a range of solutions, they considered their vast forests. Dr Qing Li, the president of the Japanese Society of Forest Therapy, and his colleagues, began promoting spending time in forests as a treatment for a great range of mental health challenges. What they came to discover in their research was fascinating. Individuals saw significant benefits to their stress levels, anxiety, depression, anger, and sleep![24] During this time in natural environments, they simply followed the checklist above, immersing themselves in their present experience as much as possible.

This daily time I spent in nature, away from my phone, immersing myself in more positive styles of thinking, and connecting with the environment, has really transformed my way of life. As I began falling back in love with nature, I found myself falling in love with my own true nature as well. Our instincts want us to be kind, to be connected with people, to eat nutritious foods, to move regularly, to sleep deeply, and to spend a lot of time outdoors. It almost felt as though the more time I spent in nature, the more 'natural' I wanted to become. I believe you will experience this exact transformation too.

Strategy

When looking at how you can increase your serotonin, think about two aspects:

- **START SPENDING TIME IN NATURE ALONE.** Use this time to deeply connect with the environment around you. Tune in to your senses, observe what you are looking at, smelling, and hearing. It may not immediately make you feel great – you need to spend time building a relationship with the outside. As this relationship builds, the value nature can bring to your life will build accordingly.

- **NATURE SHOULD BECOME A KEY PART OF YOUR SOCIAL LIFE.** Go for walks with your friends and family. This is imperative. Learning to socialize outdoors is a magical way you, and those you love, can deepen their connection with it and with one another.

Remember, when you are out there, focus on training your new Gratitude, Achievements, and My Pursuit habits.

Challenge

I would now like you to complete the 'Nature Challenge'. In order to complete this, you need to complete three headphone-free walks throughout the next week.

12

Harness the Power of the Sun

Rate on a scale from 1 to 10 how much time you are spending outside.

First, let's rate how much sunlight you are getting.

1 ➔ 10

1 = none
10 = tons

Understanding SUNLIGHT

A massive contributing factor to boosting serotonin is the sunlight we receive.[25] Before alarm clocks, it was the bright morning sunlight that woke us, the energizing midday sun that recharged us and the calming warm evening light that helped to put us to sleep. We once had an intricate relationship with the rise and fall of the sun each day, and we still need it now.

I'm sure you agree that when it's sunny outside you feel a little more positive and energetic. This isn't a coincidence; regular sunlight is absolutely vital for a huge number of health factors. Such areas include the quality of your sleep, the strength of your immune system, and the production of serotonin, which means greater energy and a more positive mood.[26]

Throughout this chapter, we are going to assess how much time you spend outside in sunlight. I should note here that any time you are completing your Nature Challenges from our previous chapter, you are also achieving your **Sunlight** goals. However, there is a key reason that nature and sunlight are separate. There may be certain days where time doesn't allow you a nice walk in nature. But sunlight is non-negotiable. Whether that means sitting out on your patio for a few minutes (even in winter!), or having your morning coffee outside, we need to be sure you are regularly immersing yourself in sunlight. And it doesn't need to be a fine day for you to experience the desired rise in serotonin. When it is sunny, you will, of course, experience a greater rise in serotonin, but it is on the cloudy days that it is even more important to prioritize getting outdoors. This is because we can all struggle in the darker winter months with Seasonal Affective Disorder (SAD),[27] due to a huge reduction in the amount of sunlight we receive each day. One study found that serotonin levels are lowest in winter, so whether it is sunny or cloudy, our time spent outside is crucial for a happy, healthy mind.[28]

The circadian rhythm

The first thing we must learn to understand is our circadian rhythm. This refers to the physical, mental, and behavioural changes that occur during a twenty-four-hour period[29] as a result of the changes in light throughout the day.[30] Your circadian rhythm impacts a range of functions within your brain and body, including your alertness, tiredness, appetite, and body temperature.[31]

Think of your circadian rhythm as your body's internal clock, always running in the background with the intention of optimizing your energy levels and mood. You want to be very alert and awake throughout the daylight hours and deeply asleep throughout the night hours in order to recharge. For our ancestors it was very easy to have a healthy and natural circadian rhythm. They woke up to sunrise, and they fell asleep after the sun set. For us, however, things are different due to the abundance of artificial light. We are no longer forced to align with the laws of nature. This creates a great range of challenges for our health, particularly a lack of natural sunlight early in the morning and seeing too much artificial light late into the evening.

In a meta-analysis (a study of lots of other studies), over 85,000 research papers have concluded that 'avoiding light at night and seeking light during the day is a simple and effective method to improving mental health'.[32] We are now going to work out precisely how much sunlight you need each day and the steps required to achieve it.

STEP 1:
MORNING SUNRISE

Bright, white, morning sunlight doesn't just activate your serotonin but it also provides a healthy, natural rise in your cortisol and dopamine.[33] This sunlight is essential to starting your day in an energetic, positive, and motivated mood. Importantly, it does not need to be sunny for you to experience this increase.

You will need:

- **SUNNY DAYS: 5–10 minutes**
- **CLOUDY DAYS: 10–15 minutes**
- **DARK, OVERCAST DAYS: Up to 30 minutes**

If it's sunny, put your sunglasses on after ten minutes (your brain needs to register the daylight unfiltered). When you are outside, ensure you spend time looking towards the sun. Of course, and this is very important, do not stare directly at the sun in a way that causes any pain or damage to your eyes. Your goal is to simply face towards it in a way that feels comfortable and safe. Think about the sun as a wireless charger – the sun has the capacity to give you energy for the day ahead.

During the winter months, or if your job requires you to wake while it is still dark, I suggest you immediately turn on a range of lights upon waking. This will contribute to starting this process. Then, once the opportunity presents itself, grab ten to fifteen minutes outside.

STEP 2:
LUNCHTIME SUNLIGHT

Your lunch break is another great opportunity to receive natural light to your eyes and on your skin, while taking care not to let your skin burn. It is very easy to spend a huge proportion of your day inside. Many of us struggle with afternoon crashes, and sunlight around midday provides a key solution to this. During your lunch break, grab fifteen to twenty minutes outside, alone or with a friend, to recharge your brain and body for the afternoon, avoiding the mid-afternoon slump.

STEP 3:
EVENING SUNSET

In the evening, warm light is incredible for relaxation and for promoting good-quality sleep.[34] Your brain is designed to associate this warm sunset light with feeling calm and sleeping deeply. An important additional brain chemical to understand here is melatonin. Melatonin is the chemical responsible for creating feelings of sleepiness and ultimately putting you to sleep at night.[35] Interestingly, serotonin is the precursor to melatonin, meaning serotonin supports the construction of this chemical.[36] With this in mind, the more serotonin you can create throughout the day, and particularly in the evening through observing the sunset, the faster and the deeper you will sleep at night.[37]

Strategy

Your strategy here is to ensure you start experiencing sunlight at three key moments throughout your day. First thing in the morning, once around your lunch break, and once in the evening. The minimum amount of sunlight you need per day is sixty minutes.

In the winter months, this becomes progressively more challenging as daylight hours can be short. This creates a range of challenges for our brains. Focus during the winter months on ensuring you are seeing sunlight during mid-morning work breaks, lunchtime, or mid-afternoon work breaks.

Remember, always pack gloves, a jacket, and a hat so the cold isn't an excuse for staying indoors. Besides, we know the cold is good for your dopamine anyway, so get yourself out there, whatever the weather!

Challenge

I would now like you to complete the **'Sunlight Challenge'**. In order to complete this challenge, focus on morning sunlight. Ensure that every day for the next week you head outside after waking and get your morning sunlight exposure. Be sure to observe your mood and energy levels as you embark on this challenge!

13

Good Health Is Made in Your Gut!

First, let's rate how healthily you feel you are eating.

Rate on a scale from 1 to 10 how healthy your nutrition is.

1 → 10

1 = very unhealthy
10 = very healthy

Understanding GUT HEALTH

The world of nutrition is a vast one, and one with a huge range of opinions, both ethically and scientifically. Whether you eat a more 'traditional' diet of meat, fish, vegetables, and fruit, or a vegetarian or vegan diet, the key components of the guidance below will be the same.

First, we now know that we need to get **Gut Health** right, as up to 95 per cent of serotonin is created in the gut[38] and we know the food we eat directly impacts our mood.[39] We now face many challenges when it comes to what we eat. There are delicious, highly calorific foods everywhere we turn. Whether we're in a supermarket, walking down the street, filling up our car at the petrol station, or eating in a restaurant, there is tempting food everywhere. We are veering further and further away from the natural foods that have been available to us for hundreds of thousands of years and closer to a highly processed, unnatural way of eating. These 'ultra-processed foods' have been proven to have an extremely negative impact on our brains and our bodies.[40]

How do we know what is good to EAT and what we should avoid?

This is simple: we want highly nutritious foods to be entering our gut in order to create serotonin, optimize our energy levels, and improve our mood every day.[41] That doesn't mean you can't have food that is fun and sometimes indulgent. Life isn't about being perfect all the time. It is about finding a consistent, easy-to-follow path to gut health that you can easily maintain for years and years. With this in mind, I have broken a range of highly impactful nutritional strategies into six key steps. How could you adopt a few of these habits in your day-to-day life?

STEP 1:
EAT UNTIL YOU ARE 80 PER CENT FULL

Let's first consider the quantity you are eating. Food, particularly the more unhealthy, processed food, can be so insanely delicious that it is incredibly hard to stop yourself eating huge portions and eating them extremely fast, too.

It is interesting here to look at the five areas around the world that are home to the most centenarians (people who reach a hundred years of age). These places are known as the 'Blue Zones'.[42] We can study both their diets and lifestyles in order to gain insights as to why this might be. The five areas are: Ikaria in Greece, Okinawa in Japan, Loma Linda in California, Nicoya in Costa Rica, and Sardinia in Italy. One eating habit that is common to people in all these places is the concept of eating until they are '80 per cent full'. Many of us are aligned with eating until we are more like 150 per cent full, eating beyond our maximum capacity. This creates a variety of challenges for our bodies. Even if you are eating healthy, nutritious food, overeating makes you incredibly tired. If you struggle with low energy levels, you definitely need to cut down to the 80 per cent rule. If you consume a huge amount of food, your body is going to have to direct a huge amount of energy to digest that food. This will then lead to you experiencing significant energy crashes.

 CHALLENGE: Slow down the pace you eat and the quantity of food on your plate. Between each mouthful of food, ensure you chew properly (this is also great for sculpting your facial muscles[43]) and ensure you take a small breath between each bite.

STEP 2:
EAT FRUIT AS A SNACK

How often do you hear someone say, 'I've got a sugar craving' or 'I'm having sugar cravings.' It is really important to understand that this is not unusual. We have craved 'sugar' since the beginning of our evolution. Sugar is a fantastic source of quick-release energy for our body. The difficulties arise now because we have access to lots of processed sugary foods, whereas originally sugar would have come from fruit.

The impact that fruit has on your serotonin system is phenomenal. It can be one of the fastest-acting positive effects on your serotonin.[44] Fruit provides an unbelievable amount of vital nutrients to our brain and body, leading to significant improvements in our energy and mood.[45] A range of foods, including fruit, include an essential amino acid known as tryptophan[46] and it is this that provides a key building block for the production of serotonin.[47]

 CHALLENGE: If you struggle with sugar cravings and snacking, fruit is your solution. The next time you are in the supermarket, I want you to buy a few different fruits – bananas, blueberries, raspberries, mango – whatever you enjoy. This week when you are in a moment of craving unhealthy food, eat fruit first. This will satisfy your sugar cravings and they will fade away. Not only is this reducing your processed sugar intake of crisps, chocolate, sweets, or biscuits, but you will also be adding incredible nutrients to your gut, enhancing your mood and energy.[48]

Tip: include a source of fat with your fruit, for example good-quality natural yoghurt and/or some nuts. This source of fat, along with the sugar from the fruit, will lead to a balanced glucose (sugar) level in your bloodstream, which will further optimize your energy.[49]

STEP 3:
INCREASE YOUR PROTEIN INTAKE

Protein is one of the most important aspects of your nutrition. The word 'protein' comes from the Greek 'proteus', meaning primary,[50] and protein has always been the primary food we have searched and hunted for. Protein is required in order to build and maintain your body muscle, and it impacts your satiety (how full you feel), which is going to help enormously with your first step, eating until you are 80 per cent full.

Important note: protein is a key building block of tyrosine, the primary amino acid building dopamine. So this is supporting your motivation too!

Take a moment now to imagine two scenarios. In one scenario I hand you a plate of food that is very heavy on carbohydrates; something like white pasta, white bread, potatoes, or chips. If I told you to eat as much as you could of this food, you would likely discover you could eat a huge quantity. Alternatively, I hand you a plate of food that is very protein dominant, something like chicken breast, salmon, steak, eggs, or tofu. In this situation if I asked you to eat as much as you could, you would likely discover that you fill up much quicker. This is because protein and the key amino acids that are in it satiate your body in a much more efficient way.

With this in mind, if you are someone who struggles with portion control and extreme hunger, a fantastic solution is significantly increasing the amount of protein in your diet. If you are wondering how much protein is a healthy amount to be consuming, you should consume one gram of protein for every pound that you weigh.[51] For example, if you weigh 150 pounds, you should aim to consume 150 grams of protein per day. You may well discover that in

Recommended sources of PROTEIN include:

TRADITIONAL DIET		VEGAN / VEGETARIAN	
MEAT	FISH	DAIRY	PLANTS
Chicken Grass-fed beef Turkey	Salmon Trout Cod	Yoghurt	Tofu Seitan Beans and legumes

order to achieve this you need to consume significantly more protein than currently.

 CHALLENGE: Throughout the next week, eat meals that include far more protein. Consider making protein the primary form of food on your plate, rather than carbohydrates such as pasta. Check out the next step, making vegetables your carb source, alongside this.

STEP 4:
PRIORITIZE VEG AS YOUR CARB SOURCE

An additional challenge with our modern diet, affecting both our weight management and our energy levels, is our often massive consumption of carbohydrates. I want to be very clear here, I am not demonizing carbohydrates. I believe a balanced range of protein, carbohydrates, and fats is the healthiest, most sustainable way to eat. There are, however, a range of carbohydrates that provide significant spikes in our sugar levels and cause crashes to our energy levels and mood.

The kind of carbohydrates that cause these spikes and crashes include white bread, white pasta, cereal, crisps, and fries. These are known as 'simple carbs'. They contain low levels of fibre, meaning there is nothing to slow down their absorption into the body, and they have a large amount of starch, which means the body quickly converts them into sugar. Overconsuming these carbohydrates will hinder your ability to sustain a healthy body weight and experience consistently high levels of energy.[52]

Carbohydrates are a vital source of energy for your body, so you do need them in your diet. Now that you are consuming fruits when snacking and adding yoghurt and nuts, you are already enjoying a good amount of carbohydrates. In addition to this, a fantastic way to feel satisfied after a meal is making vegetables the primary source of your carbohydrates. Vegetables such as courgettes (zucchini), aubergines (eggplant), carrots, broccoli, and peppers are considered 'complex carbs'. This means they contain a high level of fibre and the nutrients are absorbed slowly into your body, providing it with much more consistent energy. They also include an incredible range of vitamins and minerals, which nourish your body and optimize your gut health.

Legumes (lentils, peas, and beans) are another great choice in place of simple carbohydrates. These are another group of foods that are often consumed in the Blue Zones (see page 192). They are very fibrous and provide a slow and steady release of energy into your body. If you eat a more plant-based diet, legumes are also very important in ensuring your body is consuming a healthy quantity of daily protein.[53]

 CHALLENGE: Throughout the next week, experiment with having a good protein source on your plate, alongside vegetables as the primary form of carbohydrates. Your meals will satisfy you more and provide far more nutrients to your gut.

STEP 5:
REMOVE ULTRA-PROCESSED FOODS

Ultra-processed foods, or UPFs, are industrially made and are intentionally designed to be highly addictive.[54] Examples of UPFs include crisps, cereals, fast food, chocolate, sweets, pastries, doughnuts, and processed meats, such as ham, bacon, hot dogs, and cured meats.[55] These foods have been shown in countless studies to have a very negative impact on the health of your gut (and therefore your serotonin production), leading to significant consequences to your mental and physical health.[56]

A simple way to check whether something is a UPF is to look at the ingredients list on the back of the packet. If there is a huge list of ingredients and most of them are words you have never heard of, you want to avoid that product. These ingredients are carefully designed by scientists to make food addictive. In order for you to have a healthy gut, you need to be consuming the most natural foods you can. The ideal is to buy items that contain one ingredient: the piece of fruit, the vegetable, the egg, or the fillet of fish. This is how nature intended our diet to be. These are known as 'whole foods'.

Now, of course I understand that it is nice to treat yourself from time to time with 'unhealthy' food. I enjoy this too. It is important, however, to make tactical, thoughtful decisions. Take, for example, a chocolate bar. Your typical milk chocolate will contain ingredients that make it very moreish. However, if you select a bar that is above 70 per cent cacao solids, it will contain far fewer ingredients, have a lower sugar content, and make far less of a negative impact on your gut and brain. This is about making smarter choices when you want to have a treat.

A simple dietary rule you can follow is the 80:20 rule. This means ensuring that 80 per cent of the food you are putting in your body is healthy, natural, and nutritious. You can then allocate 20 per cent for treats, which might be an occasional bit of chocolate, some ice cream, or some crisps. Of course, it would be incredible if you could eat healthily 100 per cent of the time, and if you can, please do. But if you are seeking a more sustainable, healthier lifestyle, or you are at the start of the change, try the 80:20 rule.

 CHALLENGE: Next time you are in the supermarket, use a bag to divide your trolley into two sections. Keep the 80 per cent of the trolley that is close to you full of nutritious food. Allow yourself a handful of treats in the final 20 per cent. This way, you will start coming home with far more healthy food than unhealthy food and ultimately your diet will improve. In moments of tiredness, sadness, or boredom, our willpower will often reduce, and avoiding the UPFs becomes harder. The fewer in your house in these moments, the better.

STEP 6:
EXPERIMENT WITH INTERMITTENT FASTING

Intermittent fasting (IF) is a phenomenon that has taken the world by storm. IF refers to shortening the window of time in which you eat each day. For example, if you are currently eating your breakfast at eight o'clock in the morning and eating your dinner at eight o'clock in the evening, your eating window is twelve hours long. But if you shift your first meal later in the day, towards 11 a.m. or twelve noon, this will shorten your eating window to eight or nine hours.

Note: Throughout this fasting period, it is important you continue to hydrate by drinking water.

People who eat this way often observe substantial benefits. This includes improvements in their motivation levels and ability to concentrate when working, more consistent energy levels, and their weight becomes easier to maintain.[57] A range of research studies have even begun demonstrating the significant impact this can have on slowing the speed at which our bodies age.[58]

Important note: IF is not something you have to do in order to be healthy. If it doesn't sound appealing to you, or if you try it and it doesn't feel good, then stick to your current timings. The most important aspect of your nutrition is what you are eating, not when you are eating it.

I used to be someone who woke up every morning and ate breakfast very quickly. When I originally began researching IF, I thought, 'No way, I'll be so tired, I won't be able to work.' However, I gave it a try and it was incredible how quickly my body adapted and began to thrive as a result of just delaying my first meal.

I know you might have the well-repeated sentence in your mind: 'Breakfast is the most important meal of the day.' However, it is interesting to consider that this phrase was actually an ingenious marketing campaign throughout the early twentieth century by Kellogg's in order to promote the consumption of their cereal.[59] We know from our understanding of ultra-processed foods that, in fact, cereal is not something we want to be consuming every day, so trusting this popularized concept isn't the best plan.

CHALLENGE: For those of you who would like to experiment with IF, start by delaying your first meal by a couple of hours (it needs to be at least two hours for you to experience any changes) and simply observing how you feel.

Whenever you eat it, you want to consume a highly nutritious breakfast that includes a good source of protein, fats, and complex carbohydrates. In the following section (see page 199), I will also explain what fluids are good to drink throughout the morning and before you start eating.

IF can influence women's hormonal cycle. In Dr Mindy Pelz's popular book *Fast Like a Girl*, she recommends that women avoid any form of fasting from day twenty in your cycle until the onset of menstruation.[60] If you notice negative impacts on your cycle, avoid fasting altogether.

If you have any history of eating disorders, please avoid IF.

If you have any underlying health issues then ensure you always discuss any significant lifestyle changes with a doctor prior to implementing them.

The best drinks to consume

We now have a clear understanding of the importance of consuming a natural, nutritious range of foods to optimize the health of your gut and brain. Next, let's consider the drinks that you are putting into your body. I have broken this into four key categories for you to experiment with.

1. WATER

Your body's level of hydration is essential for your serotonin levels and therefore your mood and energy,[61] your attention span, physical performance, and metabolism.[62] Ensuring you are drinking a small glass of water, for example 200ml (7 fl oz), every thirty minutes is incredibly important.[63]

In addition to this, and something the Blue Zones are also very keen on, is consuming herbal tea. Herbal tea has a number of benefits, including the detoxification (cleaning) of your blood alongside boosting your serotonin levels.[64] Drinking a cup of herbal tea, for example mint, jasmine, lemon and ginger, or camomile, on an empty stomach first thing in the morning, is a very healthy way to start your day.

2. COFFEE

Caffeine is a powerful substance that can have positive effects on the brain, but can also create a number of negative challenges too. If you aren't currently drinking caffeine and your energy levels feel good, then you should avoid engaging with it. If you struggle with anxiety or stress, it's best to avoid caffeine altogether. For those of you who do drink caffeine, like myself, there are a few rules to keep in mind.

We are going to specifically look at coffee here. I recommend avoiding energy drinks, as they contain a huge range of unnatural processed chemicals as well as caffeine. Regarding tea, it does have caffeine in it, but in a less concentrated quantity, so coffee is really the one we want to optimize. Consider the timing of your first cup of coffee. Our aim here is to delay this first cup. This is one of the DOSE strategies that many people say has had the most significant impact on them. Many of us have coffee within the first thirty minutes of waking up. This causes a number of challenges for your brain and the body's energy cycle. In the morning when you wake (and especially if you get yourself outside into natural light), your body experiences a natural rise in the hormone cortisol. This kicks off your energy system for the day. If during this period we consume caffeine, this natural cortisol increase gets disrupted and the body uses the caffeine for energy instead. This is a primary cause of afternoon energy crashes.[65]

Instead, you should aim to have your coffee at least ninety minutes after waking, and preferably more like two hours after you wake up. This will provide a far more natural energy curve. I have my coffee at 10 a.m. and always engage with a challenging Flow State activity (see page 45) immediately afterwards, as the caffeine also provides a rise in dopamine to enhance concentration and productivity.[66]

3. PROBIOTICS

Another great way to enhance your gut health and your daily serotonin production is with probiotics.[67] Your gut is full of bacteria that arrive in all the different foods and drinks you consume. In recent years, we have discovered a range of bacteria that are particularly good at promoting the health of your gut, including one called lactobacillus. One study revealed the significant positive impacts that probiotic products can have on the health of one's gut and brain.[68]

A healthy gut helps produce more natural serotonin and thereby leads to a happier brain. Probiotics will provide substantial benefits to your mood, energy and immune system.[69]

> **Probiotics are not a substitute for medication. If you are taking medication as a treatment to improve your mental health, and it is working for you, then that is great, and following your psychiatrist's or GP's guidance is the key. In such instances, probiotics may provide additional support for your mind and body.**

There are a variety of probiotic drinks. A very popular option is kombucha, a drink derived from fermented green tea that tastes great and provides incredible gut health benefits. Be sure to avoid any with a high sugar content. The second is kefir, which is a fermented drinkable yoghurt. As well as the drinks, fermented vegetables such as sauerkraut or kimchi provide an additional way to incorporate these probiotics into your diet each day.

4. ALCOHOL-FREE

Alcohol has become so interconnected with many of our daily activities and social experiences, but it does provide a range of challenges for our brains and bodies. Not only does it cause huge spikes and crashes in our dopamine, causing low motivation and depressive symptoms,[70] but it also leads to significant reductions in serotonin levels as it is incredibly harmful to your gut.[71] This causes low mood, anxiety, and exhaustion.[72] I do believe our society would be in a far better position if alcohol didn't exist at all, but the reality is it does, so discovering the relationship with it that works for you is important.

For some people, alcohol is almost too pleasurable, and one glass always leads to four or five – and challenges with their mental and physical health arise as a result. Others may find having just one glass of wine or a beer far more manageable. The key here is that your brain and body are going to significantly struggle if you have more than one or two drinks in an evening.

CHALLENGE: See how much of an impact alcohol is having on your mental health by having a few alcohol-free weeks. Set yourself a goal of fourteen days, buy yourself some alcohol-free alternatives and some kombucha, and simply observe how you feel. I could not believe how different my brain and body felt when I didn't consume alcohol. Now, with a lot of intentional self-awareness, I no longer have multiple drinks and can enjoy drinking occasionally. It is worth creating a healthy relationship with alcohol as it will have a significant impact on your life.

Strategy

Throughout this chapter, you have learned a great range of strategies that will lead to you having a healthier gut, one that produces more serotonin and provides a nice balanced level of energy and positivity in your mind throughout the day.

I want you now to take a moment to consider which of the strategies for both food and drinks you feel you can action over this coming week.

Select one food strategy to action:

1. Eat until you are 80 per cent full
2. Eat fruit as snacks
3. Increase your protein intake
4. Prioritize veg as your carb source
5. Remove ultra-processed foods
6. Experiment with intermittent fasting

Select one drinks strategy to action:

1. Drink water or herbal tea every thirty minutes
2. Time your coffee carefully
3. Try probiotics
4. Go alcohol-free

Challenge

I would now like you to complete the **'Gut Health Challenge'**. In order to complete this, I highly recommend increasing the amount of protein in your diet, alongside eating fruit as snacks. Of course, there is a range of strategies listed on the previous page, so please feel free to select a challenge that feels the most valuable to you.

14

Slow Your Body, Slow Your Thoughts

SEROTONIN
UNDERTHINKING
SEROTONIN
UNDERTHINKING
SEROTONIN
UNDERTHINKING
SEROTONIN
UNDERTHINKING
SEROTONIN
UNDERTHINKING
SEROTONIN
UNDERTHINKING
SEROTONIN
UNDERTHINKING

First, let's rate how much you struggle with overthinking.

Rate yourself on a scale from 1 to 10 for how regularly you overthink.

1 ➡ 10

1 = never
10 = all the time

Understanding
UNDERTHINKING

This is a chapter I have been incredibly excited to write. It is an area that I feel deserves attention and understanding. Here we are going to explore your thinking, and specifically the rather challenging experience of 'overthinking'.

Overthinking can be defined as repetitive, worrisome thoughts over key challenges or fears in your life. Throughout my career in psychology, this is a problem that comes up time and time again – whether it is when I am delivering training or via social media. So many of us get stuck in our thoughts, worrying on a regular basis and sometimes even experiencing negative spiralling and panicking. Negative spiralling refers to the experience of quickly thinking of the 'worst-case scenario', when you find your thoughts going towards the most negative outcomes. Panicking or panic attacks can also arise from this.

Our mission in this chapter is clear: to learn what causes our brains and bodies to do this and, most importantly, learn how to calm them down utilizing what I call **Underthinking**. As we dive into learning how to achieve this, we must return our attention to the vagus nerve (see page 166). Your vagus nerve is a very sophisticated mechanism that connects your brain and body, and it plays a key role in the functioning of the nervous system. Your nervous system has the capacity to energize you quickly and increase your alertness, as well as calm and slow down the pace of your mind and body.

The component that increases your alertness is known as the sympathetic nervous system. The component that calms you is known as the parasympathetic nervous system.[73] The vagus nerve plays a key role in how this system operates via a huge number of electronic nerves that are placed within key areas of your body such as your throat, lungs, stomach, and intestines.[74] It is constantly reading and assessing how calm or alert your body is and feeding this information back to your brain, and can slow down your heart rate, particularly through the way in which you breathe.[75]

At this point I need to explain why slowing down your heart rate is something that is so crucial to calming your mind and specifically your thoughts. If we journey back to our ancestors, in moments of physical danger it was essential we developed a mechanism that could increase our likelihood of survival. We needed to get a lot of blood around our body very quickly in order to increase the activation of our muscles so that we could run or fight in order to survive. The fastest way to increase the quantity of available blood within your muscles is to increase your heart rate. Simply, a faster-beating heart will circulate more blood. In order to increase the speed at which your heart beats, your body will increase the pace at which it is breathing too, bringing more oxygen into your body to oxygenate your muscles in preparation for the threat. In addition to your heart and breathing increasing in speed, your thoughts move quicker too, to identify all the ways in which your life could be taken with the intention of finding a solution to enable survival. This is your sympathetic nervous system in action.[76]

Once the threat has passed, the body then needs to go through the opposite process of slowing everything back down using your parasympathetic nervous system.[77] The quickest way to achieve this is through altering your breathing. Rather than taking short sharp breaths in, the body starts taking big slow breaths out. This sends key signals to your vagus nerve that you are no longer in danger and that your body is safe to calm itself.

Now let's consider the link between overthinking, anxiety, and your breathing. Although in our modern world you're no longer likely to be chased by a bear through a forest, during moments of fearful, worrisome thinking, your brain and body don't know that. In these moments, a similar activation of your sympathetic nervous system occurs, speeding everything up. Our mission then is to activate the parasympathetic system in the most efficient way possible. There is a fascinating area of research that supports this, known as 'vagal tone', which refers to your ability to control your body's and brain's level of alertness and fear, or restoration and calm.[78] Vagal tone is something that can be trained and improved over time through intentionally learning how to control your heart rate with your breathing.[79] Individuals who present higher vagal tone as a result of such training exhibit higher rates of serotonin, our primary focus here in Part 3![80]

Breathing strategies

It is important you learn to connect with your body and calm it down through breathing. There are two key breathing strategies that will support you in this. They can be used in acute moments of fearful and worrisome thinking, and as a daily skill that you can train yourself in each morning. This will lead to a progressively calmer mood each day.

1. RESONANCE BREATHING

This refers to slowing our breathing down to just six full breaths per minute. This style of breathing has been shown to have an incredibly positive effect on calming your nervous system, increasing your vagal tone and boosting your mood.[81]

In order to reduce the pace of your breathing to just six breaths per minute, you must break your breaths into ten-second chunks. To do this, breathe in through your nose for four seconds and out through your mouth for six seconds. You then repeat this for a few minutes. Remember, to calm our brain and body down, our mission is to extend our exhale, so we must make sure there are longer breaths out.

CHALLENGE: Start practising this skill every morning for a few minutes. Find a calming spot in your home or a nice bench on a walk in nature, and each morning sit down and begin your resonance breathing training. This is like training a muscle in the gym; the more you do it, the better you will get at it, and the more efficiently you will then be able to calm yourself down in moments of overthinking and worry.

CHALLENGE: Take a minute to try this. Put your book down. Sit in a comfortable upright position on a sofa or chair. Close your eyes. Start breathing in through your nose and out through your mouth. Once you feel settled in, begin counting. Four seconds in, six seconds out.

2. SIGH BREATHING

Another fascinating breathing strategy shown to have the calming effects we are seeking here is Physiological Sigh Breathing.[82] This is a natural process that your body goes through when it is shifting you into a state of sleep. Take a deep breath in through your nose, then once you feel 'full', you take an additional short sharp breath in, again through your nose. This is then followed by a long 'sigh' out through your mouth. So, double inhale, followed by a big exhale.

The inhalation, followed by a short second inhalation, provides a greater expansion of your lungs, causing your body to release more carbon dioxide on the exhale and therefore providing the calming impact we desire.

This is actually something you may see children doing naturally after they have felt emotional or stressed. As they are calming down you will see they take a double breath in and then a slow breath out. This is the body utilising the physiological sigh to calm the nervous system.

 CHALLENGE: Try this now. Sit in a comfortable position on your sofa, close your eyes, and start with a few inhales and exhales. Then when you are ready, perform ten 'sigh breaths'. One sigh breath equals a double inhale followed by an exhale. The bigger and louder the exhale, the better.

Your daily calming practice

To incorporate these strategies into my life, and truly calm my mind, I created a short, achievable morning practice that would enable me to practise these skills. I highly recommend you do this too. Remember your aim here is to become in tune with your body and your heart. Learning to feel it, listen to it, and ultimately calm it. With this in mind, I am going to add one additional component to your morning breathing training.

When you sit down, follow these steps:

1. Take three full inhales and exhales.
2. On the third exhale, close your eyes.
3. Take three more inhales and exhales.
4. Begin your breathing practice of choice (whichever one you feel calms you more, resonance breathing or sigh breathing).
5. Breathe in this way for two to three minutes.
6. Now connect with your body. Scan your body from head to toe and see if you can feel any sensations.
7. Start in your head. Can you feel any sensations in your eyes, nose, or mouth?
8. Then your upper body. Can you feel any sensations in your throat, shoulders, chest, or stomach? In your lower body, can you feel any sensations in your thighs, bum, or feet?
9. Once you have done your breathing and scanned your body, open your eyes.

Doing this practice each morning will take you less than five minutes and it will change your life. Your brain will become calmer, clearer, and more focused.

What to do when you are overthinking

Now, with a clear understanding of the importance of training and utilizing your breathing, I want to bring you towards two additional steps that you can use during moments of overthinking.

1. VOCALIZE YOUR THOUGHTS

First, calm your brain and body down with your breathing. If you still feel the thoughts are persisting, call, voice note, or meet up with someone you trust and feel connected with. Explain that you are overthinking and describe the situation to them. This process of explaining will help you to process it, rationalize it, and open the door for guidance. Another way in which you could achieve this is through journalling. Journalling is a wonderful process grounded in a great deal of science.[83] Simply taking out a piece of paper and writing your thoughts and feelings down will support the processing and acceptance of the worrisome thoughts within your mind.

2. TURN TO GRATITUDE

We learned a great deal about Gratitude in Chapter 9 and during times of worrisome thinking it is a fantastic tool. During these challenging moments, your brain will likely go towards fear – fear about your job, health, home, friends, family, people's opinions, whatever it may be. In a similar way that the slow-exhale breathing calms your fear response, Gratitude can do the same. Gratitude reminds your brain of all the things that are okay in your life and are providing you with stability and happiness. During moments of overthinking, this offers your mind the reassurance it needs to settle down.

Becoming an underthinker is possible, starting from today.

Strategy

Throughout this chapter, you have learned about three primary strategies that you can utilize in order to calm an overthinking mind. Examples of this are:

> 1. **SLOW BREATHING**
> 2. **VOCALIZING YOUR THOUGHTS**
> 3. **TURNING TO GRATITUDE**

In order to incorporate these strategies into your life, it is important you make them a daily exercise. Building the breathing practice into your morning routine is key. Select a moment you can commit to every day. For example, on my morning walk I have a particular bench where I do my breathing. If you prefer being in the peace of your home, select a calm spot to complete your morning practice. Ensure it is always at the same time for optimum habit formation, for example after your shower, once you have put your clothes on, or after you have brushed your teeth.

Begin experimenting with vocalizing your thoughts to people you trust, alongside turning to gratitude in a moment of overthinking. Remember, these skills take practice, and the more often you engage with them, the calmer your mind will become.

Challenge

I would now like you to complete the '**Underthinking Challenge**'. In order to complete this, you will complete a short, two- to five-minute calm breathing practice every morning for the next week. Carefully observe how this impacts the peace of your mind. Remember, training your breathing is a skill. The slower the exhale, the calmer your thoughts will become.

15

Recharge at Night to Thrive by Day

First, let's rate how well you are sleeping.

Rate yourself on a scale from 1 to 10 for the quality of your sleep each night.

1 → 10

1 = terrible
10 = incredible

Understanding DEEP SLEEP

Throughout Part 3, we have been discovering the key behaviours that optimize the health of your body and, as a result, the health of your brain. You should now have a clear idea of the kind of activities you can do throughout your day to increase your serotonin levels. This includes early morning sunlight, time in nature, eating nutritious foods, and calm breathing patterns. The final component to serotonin comes at the end of your day, your sleep.

In this chapter, we are going to learn why the quality and quantity of your sleep are absolutely essential to improving how you feel in your mind and, most importantly, how you can improve your sleep, every single night.

Deep Sleep leads to huge increases in your serotonin levels.[84] Given that we know that the function of serotonin is to improve your mood and energy levels, you can see how interconnected your sleep is with this. I am sure there have been many days when you have underslept and felt a significant dip in your energy and your mood the next day. Often increased irritability is a key sign that you have underslept.

Good-quality sleep will not only boost serotonin but also impact a number of additional key factors such as your memory, your attention span, your learning efficiency, your emotional processing, and even your metabolism and capacity to maintain a healthy body weight.[85] It is important to understand that sleep is a big topic with a significant number of influences that will impact it.

NOTE: The quality of your sleep is more important than the quantity. As a rough guide, you should aim for between seven and nine hours of sleep per night. The best way to assess your sleep is simply by observing how energized and alert you feel when you wake up.

Optimizing your sleep

Here are a number of actions you can take that will optimize your sleep, from your ability to fall asleep quickly to achieving the deep states of sleep that are required to recharge your brain and body. You don't need to do everything I guide you through here. All are important, but just select a few that feel manageable and useful for you in your life right now.

1. MORNING SUNLIGHT

As we discovered in Chapter 12, Sunlight, the time when you view light is crucial for optimizing your circadian rhythm. To remind you, this is your internal body clock and it's responsible for waking you and putting you to sleep. One of the most important components of improving our sleep is reducing the time it takes from when we wake up to when we get outside and see daylight for the first time. The moment your eyes are in daylight, the waking component of your body clock starts. The faster your energy system starts in the morning through sunlight, the faster it will calm in the evening, and with that, the more efficiently you will fall asleep. Aim to view sunlight within a maximum of thirty minutes of waking. View it for a minimum of ten minutes.

In order to wake up with ease through the darker winter months, I highly recommend considering a 'sunrise alarm clock'. These wake you with light throughout a thirty-minute period, providing a far more natural waking cycle for your brain and body. Getting one genuinely changed my life.

2. DAILY MOVEMENT

Now, with you awake and on the move, it is important to think about how much movement you are actually getting every day. If you are someone who lies in bed struggling to fall asleep or you wake in the night, this is a crucial part of your life to consider. Very simply, your body needs to physically require sleep in order to desire it. If, and it is very easily done in our modern world, you live a pretty sedentary life, navigating between your bed, car, desk, and sofa, and don't physically push your body, then your body won't need the sleep.[86] Ensuring that you are moving your body each day is a non-negotiable. In the first chapter of Part 4, Endorphins, we will be exploring how you can get more daily movement into your life.

> **If you are currently navigating an injury or have a disability that doesn't enable movement, use one of the actions on page 222 to settle your mind.**

3. YOUR ENVIRONMENT

There are a few crucial environmental components that will help you to achieve optimal sleep.

- **Temperature:** Your body reduces by one degree Celsius as it falls asleep[87] and it is therefore important that your bedroom is cool. I leave my bedroom window open throughout the night in the summer and in the winter I leave it open for twenty minutes just before I go to bed.

- **Lighting:** Just as we need bright light in the morning to wake our body, we need dim lighting in the evening to calm us. It is really important that after 7 p.m. you never have any main lights on in your home and use lamps instead. The aim is to avoid having any light that is coming from above your head (your brain perceives this as the sun). Ensure all lighting is coming from below eye level.

- **Comfort:** It is important you are comfortable at night. Whether this means getting a new pillow, a new duvet, or even a new mattress, comfort is essential. Cleaning your bedding on a regular basis and treating your bedroom with care is key. Create a calming sanctuary for your sleep. Keeping the space organized will also help optimize your dopamine.

4. NIGHT-TIME TECHNOLOGY

How you interact with your tech in the evenings and before sleep is crucial.

- **The type of content:** In the last thirty minutes before sleep, it is essential that you are not scrolling social media or news platforms. This is far too stimulating for your brain and either leads to an inability to fall asleep or increased waking in the night.[88] Reading is optimal, as it is incredibly calming for your mind. For example, consider reading a few pages of this book before going to sleep. If there are evenings where you don't want to read, you are much better off watching a TV show or listening to a podcast over scrolling.

Note: When looking at any screen after 7 p.m., ensure the brightness is extremely low. Bright lights will wake your brain.

- **Charging your phone:** You may have spent the last ten years charging your phone right by your head. There is a really important reason this must end today. Research has shown that viewing phone screens between 11 p.m. and 4 a.m. leads to an overactivation of a part of your brain called the habenula.[89] Overactivation of this area is commonly associated with clinical anxiety and depression. This means that even if things are going well in your life, simply viewing your phone in the night could be leading to depressed and anxious feelings. Charging your phone either on the other side of the room or preferably in another room (I charge mine in the lounge) is essential. This will also support your Phone Fasting, as it is crucial you don't get your first hit of dopamine from your phone each day.

- **If you fear emergency calls**, you can have the favourites in your contacts able to ring you even when your phone is on do not disturb. In addition, if emergencies are your main argument against charging your phone away from your bedroom, ask yourself how many emergency calls you have actually had in the night. Ensure your phone is on do not disturb from 8 p.m. to 8 a.m.

5. YOUR DIET

There are three crucial items you may be consuming in the evening that are ruining your sleep. This may be annoying to hear, but they are caffeine, alcohol, and sugar. I understand these substances can provide a great deal of pleasure, but the negative impact on your sleep and your mental health is worse than any pleasure they may be giving. It's not about cutting these products out, it is about making smart decisions.

- **Caffeine:**[90] If timed correctly (consumed a minimum of ninety minutes after waking), caffeine can support your energy and capacity to enter Flow State. It is important that this is your only coffee and you don't have any more caffeine after 12 p.m. (this includes tea). Opt for decaffeinated or herbal tea in the afternoon.

- **Sugar:**[91] Eating sugar in general is problematic. But eating sugar after 8 p.m. also destroys the quality of our sleep. It leads to you waking in the night. Ensure that in the evening you are disciplined with avoiding sugary treats. If you need a treat after dinner, opt for natural sugars such as yoghurt with good-quality honey, fruit, or occasionally a couple of squares of dark chocolate (70 per cent or more cacao solids).

- **Alcohol:**[92] I totally appreciate that alcohol can become a huge part of your evening routine, and many of us actually think it helps us to sleep. In fact, it has the opposite effect. Alcohol stresses your body, waking it more in the night and significantly reducing the deeper stages of sleep that are needed to recharge your brain. On weekdays, opt for a herbal tea or even a kombucha (which contains minimal caffeine) in the evening and avoid alcohol. Observe your sleep, mood, and energy levels, and decide for yourself whether this is a good decision.

6. CALMING YOUR MIND

Finally, let's look at some science-backed methods you could use to calm a busy mind when trying to fall asleep.

- **The consistency of your bedtime:**[93] Your brain is very good at learning. Having a consistent bedtime every day leads your mind to instinctively prepare itself for sleep.

- **Gratitude:**[94] During Part 2, we learned about the benefits of Gratitude. This is a great way to begin calming your mind before sleep and when you are lying in bed, particularly if you feel worried at night.

- **Underthinking:**[95] After going through your gratitude practice, begin your resonance breathing (see page 209). This will slow your heart rate and improve how quickly you can fall asleep. Ensure you do this when waking in the night to calm your mind back down.

- **Writing:** If you struggle with lots of thoughts either before sleep or during the night, then writing is a great option.[96] Have a pen and paper by your bed and in moments of overthinking write your thoughts down. This will help to get them out of your mind, and enable you to fall back to sleep.

- **Listening:** On the nights that you really can't sleep, don't fight it, accept it, and just ensure your body is in the calmest position possible. Listen to 'sleep music', which may include a calming sound of nature or a gentle beat known to calm our brains. Alternatively, choose a relaxing podcast, as this will settle your thoughts.[97] Consider using a tablet that has no social media or email on it. Let yourself enjoy what you are listening to.

Strategy

I want you to take a moment to look through these key strategies and consider which is the most important for you to implement right now. Please remember, you don't need to do them all, just the ones that instinctively feel of value and achievable.

1. MORNING SUNLIGHT
2. DAILY MOVEMENT
3. YOUR ENVIRONMENT
4. NIGHT-TIME TECHNOLOGY
5. YOUR DIET
6. CALMING YOUR MIND DOWN

Challenge

I would now like you to complete the '**Deep Sleep Challenge**'. In order to complete this, you will need to make your sleep a priority throughout this week. Implement your key strategy of choice.

Building Your DOSE

Throughout Part 3, we have explored the true power of boosting your serotonin levels to create a happier, more energetic life.

You have been on an adventure of experimenting with a range of new behaviours, which will support optimizing this system in your brain and body.

I now want you to take some time to consider which one of the five primary serotonin actions feels the most important for you to continue to prioritize. It would be incredible if all of these behaviours remain a priority in your life. However, selecting one core action and ensuring it becomes deeply embedded in your life is essential.

WHAT WILL BE YOUR PRIMARY
Serotonin Action?

1. NATURE
Regularly connecting with the natural world, headphone free

2. SUNLIGHT
A daily priority to view sunlight as quickly as possible after waking

3. GUT HEALTH
Scrapping processed foods and returning to a whole-food diet for your body

4. UNDERTHINKING
A daily practice to calm your body and therefore calm your thinking

5. DEEP SLEEP
Optimizing your sleep to optimize your wellbeing

Be sure to chat with a friend or family member about the primary serotonin challenge you have selected!

PART 4

Destress and Calm your Mind

Understanding ENDORPHINS

Welcome to Part 4 of your DOSE journey: Endorphins. Endorphins are chemicals within your brain and body that have an exceptional ability to help you navigate stress and improve your physical health, two things I imagine you likely desire. Learning to understand their function in your life and how you can increase their activation on a daily basis will be transformational. Throughout Part 4, we will dive into the reason this mechanism evolved and, importantly, a fun and engaging range of challenges designed to boost your endorphins.

In order to understand your endorphins, I need to return to our ancestors and how endorphins would have saved their lives on a regular basis. Imagine you are a hunter-gatherer, traversing a challenging terrain. You are tired and hungry. Suddenly a predator appears and your life is in significant danger. Your brain and body are faced with a choice – run from this animal or attempt to fight it. You choose to turn and run. Stress is rising in your mind, you run as fast you can, a strong pain begins within your stomach (now often referred to as a stitch); you ignore the pain, you keep running and running. The pain begins to subside as you run for longer. You remain focused on one thing, survival. After ten minutes, you realize you are safe and can begin to recover.

During this pursuit of survival, your endorphins have played a key role. In the immediate moment of hard physical exertion, your brain and body have begun releasing endorphins for two primary reasons. Number one, to remove the stress from your mind so that you can focus on staying alive.[1] Number two, to remove any pain from your body.[2] If I asked you to run at your maximum speed for ten minutes, I'm sure you would experience pain, maybe a stitch, maybe an ache in your knees. The endorphins released in this stressful situation work as a natural painkiller to provide you with the greatest likelihood of surviving.

Now, in our ever-advancing modern world, I imagine you aren't spending your days being chased by predators. However, I'm sure stress is still something you experience regularly. Learning to intentionally activate endorphins will de-stress your mind and body, promoting feelings of relaxation.[3]

The Endorphin Principles

PRINCIPLE 1:
THEY REQUIRE HARD PHYSICAL EXERTION

Exercise is, of course, a key factor here. When you physically push your body, it provides a significant release of endorphins.[4] This is the first key principle to understand: you are required to push your body in some way. For example, you may have come across the term 'runner's high', which happens to people who run for long distances and experience a significant elevation in endorphins as a result of the physical pain their body is going through.[5] Despite this, exercise is just one of five primary methods that we will use to start activating your endorphins more regularly. There are other ways we can ensure your body is physically exerting itself.

If you are currently navigating a disability or injury and cannot exercise, we will be exploring a number of other ways you can boost your endorphins in this section. To check out a summary, see pages 290–291.

PRINCIPLE 2:
THEY ARE A NATURAL BRAIN AND BODY DE-STRESSER

The second principle I want you to associate with endorphins is simply imagining them as your brain and body's primary de-stresser. Whenever you find yourself feeling stressed out, I immediately want you to think, 'Ah, I must boost my endorphins.'

Do you have low endorphin levels?

If you regularly feel stressed out, this is a key sign your endorphins need boosting.[6]

If you struggle with feeling angry and frustrated at times, you could benefit massively by increasing the activation of endorphins.[7]

It is important to note that these emotions are completely natural. Feeling stressed and frustrated at times is part of the human experience. My focus here is helping you to understand that in the moments where you feel this way, you need to immediately think how you can boost your endorphins.

The FOUR CAUSES of low endorphins

1 **The first cause is a lack of hard physical exercise.** For 300,000 years countless generations of our human ancestors spent a significant proportion of their days on the move, hunting, foraging, building, and exploring. Much of this would have been hard work, physically pushing their bodies in order to survive. The absence of some form of exercise regime in your life will lead to a significant reduction in endorphin activation.[8]

2 **The second key cause is a sedentary lifestyle.**[9] With modern innovations, living a life that lacks much movement at all can occur very easily. We have the ability now to do so much without moving; food can be delivered to your home, work can be done on your laptop while sitting on the bed, and cars and other transport can take you everywhere you need to go. Often we undertake no proper exercise and it has been proven that a lifestyle that involves very little walking will reduce your endorphin levels.[10]

3 **Our third key cause is a lack of laughter.** Laughter is one of the primary ways in which we raise our endorphin levels, as we will come to discover in Chapter 19.[11] More and more time is being spent behind our screens, be that at work or as part of our social lives. Ask yourself: how regularly do I really laugh during these times? I imagine it is significantly less than when you are having fun, in-person social engagements. This one is not to be underestimated – there is a reason why laughter is so enjoyable. Putting yourself in environments where laughter can occur is non-negotiable.

4 **Our fourth and final cause is chronic stress.**[12] Chronic stress refers to ongoing, persistent stress that continues over an extended period of time, typically weeks, months, or in some cases, years. This is an important concept to understand and take action on. Many people feel chronically stressed, and breaking this cycle is absolutely essential. Innovative neuroscience demonstrates that repetitive overactivation of cortisol, your primary stress hormone, will result in reduced functioning of your endorphins.[13] A challenging cycle will then occur. You progressively become more stressed, while the key chemical that can de-stress your mind has a reduced ability to work, leading to the stress becoming worse and worse. It is essential that you incorporate a number of these activities into your life to break this cycle of chronic stress.

While the five endorphin boosters I describe in the following chapters are going to be the best methods to scientifically de-stress your mind and body, a number of the behaviours that you have already begun incorporating into your life will support this, too. For example, Phone Fasting (see page 65) – stepping away from your phone for periods of time – is essential and allows you to avoid your work, stressful news, or social media. Another example is the increased prioritization of your Social Life (see page 129), enabling you to connect with people and focus your mind on other people's lives.

How high endorphin levels will feel

As we work towards increasing your endorphin levels, you are going to experience two primary emotions. First, you will feel more positive in your mind[14] and a little more optimistic and happier with life. Second, you will feel significantly more relaxed.[15] You will feel calmer day to day and more capable of navigating the inevitable stresses that come up in your life.

You will find a summary on the next page of the key functions, principles, feelings, and behaviours that are associated with your endorphins.

Endorphins Summary

Function →
- Coping with stress
- Physical health

Principles →
- Requires hard physical exertion
- Natural brain and body de-stresser

Low Endorphin Symptoms →
- Stressed
- Angry

Low Endorphin Causes →
- Lack of physical exercise
- Sedentary lifestyle
- Lack of laughter
- Chronic stress

High Endorphin Symptoms →
- Positive
- Relaxed

Endorphin Actions →
- Exercise
- Heat
- Music
- Laughter
- Stretching

16

Get Moving!

First, let's rate your current physical health and fitness.

Rate yourself on a scale from 1 to 10 for how fit you are feeling now. Be honest.

1 ➜ 10

1 = very unfit
10 = very fit

I appreciate that this number will vary. Some of you reading this book may be super into exercise, and if that is the case, this chapter is going to ensure you are optimizing the way in which you are moving your body. Others may have a lower score; please know this is very normal in our society today, and working on increasing this number is going to be our focus.

Understanding EXERCISE

We know from the introduction that physical exertion is a key method of activating the endorphin system.[16] We are not going to get you running away from predators, but we want to get your body feeling as though it could. When you are exercising, it is in the moments that you really push yourself that you will experience the greatest rise of endorphins.[17] This 'push' is referring to the moment you are hiking up a hill, running as fast as you can, swimming at your top speed, cycling at full speed, or completing those last reps when you are in the gym. When your body experiences true physical pain, it will pump endorphins throughout your brain and body.

Important note: This physical pain refers to muscular and cardiovascular pain, for example your muscles burning or breathing heavily during cardio. Pain through injury or pushing yourself too hard is not the goal!

In terms of incorporating challenging but achievable **Exercise** into your life, we are going to break exercise into two key components. Both are essential to your immediate mental and physical health, alongside the longevity of your body.

Please consult with your GP before undertaking any vigorous exercise, especially if you have underlying health conditions.

Your strength

Reduced muscle mass is one of the leading causes of mortality in older age.[18]

Many studies have demonstrated that people who lose strength have an increased likelihood of falls, which lead to many of the physical and mental traumas that ultimately lead to loss of life.[19] So, it doesn't matter whether you are young, middle aged, or in your later years, strengthening your body is vital. First, let's consider how you are going to strengthen your body. The focus here is building and maintaining muscle mass, and must be intentionally incorporated into our lives. There are three key ways in which you will achieve this.

1. LIFTING WEIGHTS

Lifting weights is one of the most popular ways to strengthen your body, whether in the gym or at your home with dumbbells.

When you lift weights, you tear your muscle fibres, which then go through the process of recovery. They become stronger as they rebuild (providing you are keeping your protein intake high, a key focus of ours during your Gut Health Challenge on page 203).

Before you lift weights, always take time to warm up and stretch your body. It is also important to buy the right size of weight for you. In Chapter 20, we will cover some key stretches you can do. Ensure you train all the muscles in your body evenly: your arms, your chest, your shoulders, your back, your core, and your legs. Aim to do a minimum of two weight-lifting sessions per week. You can mix things up and break this down into an upper-body session one day and a lower-body session on another.

UPPER BODY
- **Include bench press for your chest, military press for your shoulders, bicep curls for your arms, and leg raises for your core.**
- **For each exercise, aim for <u>3 sets</u> and <u>10 reps per set</u>.**

LOWER BODY

- Include squats, lunges, and leg extensions.
- For each exercise, aim for <u>3 sets</u> and <u>10 reps per set</u>.

2. RESISTANCE BANDS

A quick search online will show you what these are, and looking for 'resistance bands workout' on YouTube will provide short routines to get you using them. These may include bicep curls, shoulder press, squats, and lunges. Find a video that feels achievable to you in terms of length and difficulty, and start from there.

Remember, the focus here is getting your body progressively stronger and boosting your endorphins in the process. In the final reps where you really push yourself, your brain will release the greatest amount of endorphins.[20]

Note: *If you have recently started training in the gym, be sure to chat to the personal trainers in order to understand the correct posture and best technique for the exercises you are doing.*

3. BODYWEIGHT TRAINING

This refers to exercises such as press-ups, pull-ups, squats, and sit-ups, which use your own body weight to build strength. Again, these are incredibly effective. Adopt the same mentality of including a minimum of two bodyweight strength-training sessions into your weekly routine.

UPPER BODY
- Include press-ups, pull-ups, tricep dips, plank.
- For each exercise, aim for <u>3 sets</u> and <u>10 reps per set</u>.

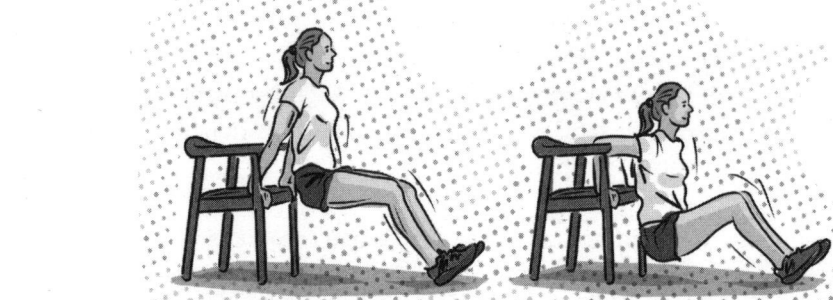

LOWER BODY
- Include squats, lunges, glute bridge, calf raise.
- For each exercise, aim for <u>3 sets</u> and <u>10 reps per set</u>.

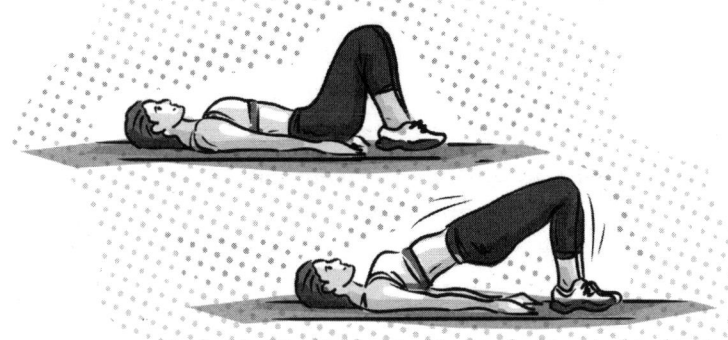

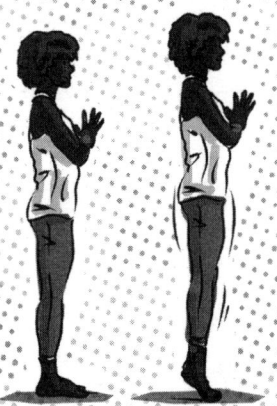

Your endurance

This refers to your physical fitness. A simple way to measure your fitness is by doing cardio for ten minutes – this could be running, cycling, swimming, or hiking – how long would it take your body to recover from this? You will notice your heart is beating quickly and you may be out of breath. The speed at which your body returns to a normal rest state is the clearest way to know if your body is physically fit. We are going to explore seven methods in order to train your endurance.

As you read through the ideas below, ask yourself which you are most likely to engage with regularly. If you are struggling to get much exercise into your life, then I want you to start small. The aim isn't to suddenly start training every day for hours, which is unsustainable. Exercise is something you gradually build up and you will naturally prefer different methods that suit you. As your body becomes fitter, it will begin to enjoy the experience of exercise more and more.

1. WALKING AND RUNNING

Running is something I have found very hard to get into. I therefore use my morning walk (when I am already getting my sunlight and nature boosts) and take three moments during this walk to push my body a little harder. This might be a fifty-metre fast jog, or a jog up a small hill. These moments provide us with a greater endorphin boost and a gradual increase in our body's physical endurance.

2. CYCLING AND SWIMMING

With both cycling and swimming, start at a difficulty level and length of time that feels achievable for you. As you gradually and successfully complete more and more sessions, the pursuit and accomplishment of these will provide a boost in your dopamine, too! As your dopamine rises, your motivation rises. Over time, motivating yourself to exercise will become easier and easier and you will be achieving the double benefit of more endorphins in your brain alongside more dopamine. Remember when you are cycling or swimming to include a few moments where you push as hard as you can. Tell yourself in these moments that you are boosting your endorphins and as a result providing the greatest stress-relieving effects on your mind.

3. GYM CLASSES AND MARTIAL ARTS

Gym classes and martial arts are fast growing in popularity. Both of these are awesome options, as you have the additional motivational component of the group around you, boosting your oxytocin.

4. SPORT

Finally, considering whether there is a sport you could play is an incredible way to achieve our goal here. For many years I let go of sport. It was something I loved in school and then gave up during adulthood. I decided to start playing tennis again and it's been amazing. The benefit of sport is you almost forget that you are exercising because you are focused on the sport itself. If there is a sport you once enjoyed, think about how you could reincorporate it into your life. Again, if it is a team sport, there is an opportunity boost your oxytocin through this too.

Whichever method or methods you select to train your body's endurance, ensure you achieve two key sessions per week. This means we are aiming for a minimum of two strength sessions and two endurance sessions in total per week. An achievable number. Remember, if you aren't feeling that fit, start with small sessions and gradually build up. If you are already smashing your exercise regime, keep at it, it will help your mind and body more than you know!

Motivating yourself to exercise

One of the challenges with exercise is, of course, motivation. To help you exercise more often, I want you to consider two key components that will encourage you to keep at it.

1. COMPETITION AND A PRIMARY GOAL

We are naturally a very competitive species and our brain is primed to chase goals (you already know this from your understanding of dopamine!). When you are working towards a specific goal, it helps your brain maintain high dopamine levels, which lead to greater motivation.

Consider signing up for a fitness challenge of some kind. If you can do this with a friend or family member, you will be even more motivated. For example, challenge your friend or partner to a competition of who can do the most steps per day in the next week. Each evening ask one another how many steps you did and see who won. This competitive element will motivate you. The same occurs if you see who is stronger in the gym, who can run faster in the park, or who can sprint-cycle at the end of your ride. Competition and clear goals are crucial to maintaining motivation.

2. SMART DECISIONS

Sedentary lifestyles are terrible for your brain and body. Alongside considering some form of fitness challenge in your life, I want you to start making smart decisions to increase movement in your average day. For example, if you drive to the supermarket to do your food shop, I want you to park at the far side of the car park, away from the supermarket, and increase your walk to it. When you are faced with taking the stairs or the escalator, take the stairs. Get up more regularly from your desk and walk around. Take short walks on your lunch breaks. Make your daily life require more moments of small movement to ensure you are subtly activating your endorphin system. This will provide a more relaxed experience in your mind and an ever improving level of physical fitness.

Strategy

Consider your method for STRENGTHENING your body:

1. LIFTING WEIGHTS
2. BODYWEIGHT TRAINING
3. RESISTANCE BANDS

Consider your method for TRAINING your endurance:

1. WALKING
2. RUNNING
3. CYCLING
4. SWIMMING
5. GYM CLASSES
6. MARTIAL ARTS
7. SPORT

Consider your method for MOTIVATING yourself:

1. COMPETITIONS
2. SMART DECISIONS

Challenge

I would now like you to complete the '**Exercise Challenge**'. In order to complete this challenge, you need to include two strength-training sessions and two endurance sessions throughout this next week.

17

Relax with Heat

ENDORPHINS
HEAT
ENDORPHINS
HEAT
ENDORPHINS
HEAT
ENDORPHINS
HEAT

Let's rate how relaxed you are currently feeling.

Rate yourself on a scale from 1 to 10 for how relaxed you feel.

1 ➜ 10

1 = very stressed
10 = very relaxed

Understanding HEAT

Have you ever got in a warm bath and felt your body immediately begin to relax? You put some candles on, you lie down in the warm bubbly water, you take a deep inhale through your nose and a slow, long sigh out of your mouth. In this moment you feel a sense of calm throughout your mind and body.

Hot environments are an incredible method you can utilize to raise your endorphin levels.[21] During this chapter, we are going to consider how often you should be using heat, how long for, and the different ways to achieve it. There are three primary methods we will explore: baths, saunas, and steam rooms.

First, let's remember how endorphins function. The primary reason your body has endorphins is that so during moments of physical' exertion your body will release them to de-stress your mind and body. When you lie down in a bath, your body temporarily experiences a form of 'heat stress'. Your body perceives that this heat could be dangerous, as it doesn't know that the water won't get hotter and burn you. To prepare for this eventuality it releases endorphins throughout your brain and body.[22] This is incredibly useful as it promotes feelings of relaxation.

It would be incredible if you could start including more baths in your daily life. When you take a bath, ensure you spend a minimum of fifteen minutes in there. Create a relaxing environment. Put some calming music on, bring *The DOSE Effect* with you, light some candles, and add some form of bubble bath or relaxation salts. This is also a perfect time to practise your Underthinking breathing (see page 209). Remember, breathe in for four seconds, out for six. Do this for a few minutes to promote further relaxation.

If you don't have access to a bath, warm showers for five minutes can help to achieve the desired effect. Baths simply provide a great immersion in the hot water and opportunity to relax.

Remember: Don't take your phone to the bath with you. Lying in the water and scrolling social media, messaging, or looking at the news will not allow your body and mind to settle down. Think of this as an additional moment to achieve a DOSE of Phone Fasting.

Note: Be careful. Never get in a bath that hurts your skin as you get in. This can be incredibly detrimental to your health. A warm, not hot, bath is what you are aiming for.

For those of you with families, you may feel as though you don't have time to prioritize yourself in this way. I want you to consider this analogy. During the safety briefing in a plane, a parent is always guided to put their oxygen mask on first before they put one on their kids, so that they have the capacity to save their kids. Think of this **Heat** self-care as putting your oxygen mask on. Taking care of yourself and de-stressing your mind only enables you to be a greater, more loving support to the people in your life.

Our second and third methods of heating the body are saunas and steam rooms. I understand that not everyone will have access to these, which is of course why we start by considering baths and showers. If, however, you can access one of these environments at the gym or spa, they are absolutely incredible for your physical and mental health.

For the optimum endorphin release, I recommend prioritizing the sauna over the steam room. Here, again, we achieve our desired aim of physically pushing your body. As the minutes go by in a sauna, you will notice it becomes more and more challenging. During these moments, in a similar way to when you push yourself when you are doing your Exercise Challenge, you will release endorphins to support your mind and body through it.

Alongside this, there are a huge range of additional health benefits that occur as a result of spending time in saunas.

1. **DETOXIFICATION:** Sweating in these environments helps detoxify your bloodstream. We are constantly consuming harmful toxins in our food, drinks, and environment, and this is a great way to expel them.[23]

2. **BOOSTING THE IMMUNE SYSTEM:** Your endorphin system and immune system are closely connected. They can increase your resilience to illness.[24]

3. **IMPROVED SLEEP:** The relaxation effect that occurs in a sauna has been shown to improve the quality of people's sleep. One study found 83.5 per cent of people saw significant improvements in their sleep.[25]

4. **MUSCLE GROWTH:** The heat stress leads to a release of your 'growth hormone', which supports maintaining and increasing your muscle mass.[26]

5. **COGNITIVE ABILITY:** In a study lasting over twenty years, more than 2,000 people were examined and it was revealed that individuals using saunas four times per week are 66 per cent less likely to be diagnosed with dementia.[27]

Progressively build up to fifteen minutes in the sauna. Of course, be careful, and if you ever feel very light-headed, then step out and take some calm slow breaths. Steam rooms are an additional way in which you can also heat your body. They are great for relaxation.

Note: If you are navigating a specific health condition such as heart challenges or high blood pressure, please check with your doctor whether it is appropriate for you.

Strategy

INCREASE the number of baths you're taking:

1. Ensure you are in there for a minimum of fifteen minutes.
2. Utilize candles, music, your *DOSE* book, bubble bath, and relaxation salts.
3. Practise your calm breathing patterns and step away from your phone.

INCREASE the number of saunas you're taking:

1. Build up to fifteen-minute sessions.
2. Don't take your phone in the sauna (*use the timer on a wall*).
3. Enter these environments four times per week.

If you are considering joining a gym, check if they have a sauna. The health benefits are enormous and gaining access to a sauna or steam room is worth it, even if it costs a little more money per month. Consider cutting back on alcoholic drinks every month to pay for the upgraded gym membership. This will give you double the benefit – less alcohol (great for your dopamine) and more saunas (great for your endorphins).

Challenge

I would now like you to complete the '**Heat Challenge**'. In order to complete this challenge, you need to ensure that you are in your chosen hot environment four times throughout the next seven days.

18

Sing Your Stress Away

First, let's rate how regularly you sing.

Rate yourself on a scale from 1 to 10 for how often you sing out loud to music.

1 → 10

1 = never
10 = all the time

Understanding MUSIC

Do you ever sing and dance to music? Have you ever found yourself driving your car, singing along to your favourite song? Do you ever sing in your home, in the shower? Have you ever been to a silent disco or a concert where everybody is singing at the top of their voices and without embarrassment? When you do this, something like euphoria can come over you and you will feel awesome. These experiences create a magical sensation in our minds, and this is governed by your endorphins.[28]

Remember, endorphins are simple: if your body physically exerts itself, endorphins will be released. In these moments of singing, of embracing **Music**, particularly if you really go for it, your body is pushed and endorphins are pumped into your brain.

Let me talk you through an experience I recently had. I had a challenging work meeting that led to my stress levels rising. I felt overwhelmed by some important decisions that needed to be made and my stress continued to rise. I couldn't stop thinking about it. My heart was racing and my thoughts were spiralling. I got in my car after the meeting to drive home. As I got in the car, I thought about this very concept, how music can create an endorphin release, and I knew doing so could de-stress my brain. I started to play some of my favourite songs. At first, in my more negative stressed state of mind, I muttered along to the songs. Gradually I got more into it, I began to sing louder (imagining I am some kind of professional, despite my questionable voice). Quickly my mind was immersed in the songs. I sang for ten or fifteen minutes. During this period, my mind let go of my worries. As I arrived home they returned to my mind. But this time I was coming at them from a different headspace, one where I felt much calmer.

In moments where our minds are stressed, solving problems is extremely hard. Music is an incredible way to temporarily disconnect from your worries and de-stress your mind, thereby enabling you to return to the problem at hand in a different frame of mind. A frame of mind that is rational and has the capacity to find your solution.

One study revealed the immediate benefits music can have on an individual's feelings of excitement about their life, seeing additional benefits of a reduction in people's anxiety and physical pain, while also improving the functioning of their immune system.[29] Another saw transformational benefits to individuals' stress response, demonstrating that those who increased the amount of time they spent engaging with music recovered faster from stress.[30] A third study saw significant reductions in the participants' cortisol (primary stress hormone). This research specifically saw improvements in people's ability to regulate their emotions.[31] Heightened levels of stress and anxiety are incredibly common for all of us these days. The fact that simply playing some music and singing along to it can in fact provide some relief from these challenges is incredible and something you can use to help wherever you may be.

It's important to note, this isn't a new phenomenon. We have instinctively known for thousands of years that singing and dancing result in positive emotions. The traditions and religious practices of many cultures have always included singing, chanting, and dancing. The human instinct is powerful; it knows what is good for it. Modern neuroscience is simply catching up and explaining why that is.

Something important to consider when getting a daily DOSE of music is the different styles you listen to and the impact this will have on your state of mind.

1. ENERGIZING MUSIC

Upbeat music that gets your heart rate going. This sort of music can be useful in the morning to get you energized for your day. It can also be great before or during your workout. If you are feeling tired, try to lift your energy by experimenting with singing and dancing to some energizing music.

2. CALMING MUSIC

Calming music is great to listen to towards the end of your day, especially when you are wanting to wind down after a busy day of work. Play some calming music during your commute or once at home. Sing along gently to this. The more you can immerse yourself in the song, the better. This provides a rest for your mind after a busy day of thinking. If you struggle to sleep, searching 'calming sleep music' on your chosen app before bedtime can also be a great way to settle your mind.

3. FOCUSED MUSIC

Some styles of music have been shown to increase your level of focus. Music that has a calm, slower beat and without lyrics can improve your capacity to focus when you are working. Experimenting with 'binaural beats' or 'Lo-Fi' music is a great idea. I find this music to be extremely beneficial when I am trying to do focused work. I select my task, close the distracting apps on my computer, get myself a nice drink, put my phone in another room, and then start my music. The ritual of this process also helps me to zone in on the task at hand.

4. SAD MUSIC

This is an interesting one. You may have noticed that the style of music you are listening to at any time can directly reflect your current mood. Sometimes when we are sad or low, our music tastes can align with these feelings. Alternatively, when we are in a happier, more excited state, our music aligns with that energy too. So if you are navigating something that is distressing, for example a breakup or losing a loved one, listening to sad music allows you to fully experience these emotions. Allow yourself to cry and feel this pain, as it can help you to process the emotion.[32]

Important note: If you get stuck in a loop of crying and feeling distressed, it can be incredibly detrimental. Going through a period of listening to sad music is okay. But ensuring you return to more uplifting music at other points in your day is important for getting your mind back into a more positive place.[33]

Music is an incredible way to temporarily disconnect from your worries and de-stress your mind, thereby enabling you to return to the problem at hand in a different frame of mind. A frame of mind that is rational and has the capacity to find your solution.

Strategy

Here our strategy is simple: we are going to intentionally start increasing how much you are singing every day. Our aim is for you to achieve a minimum of five minutes each day. Music apps have made it incredibly easy to learn the lyrics to songs now (learning the lyrics requires focus, which is also great for your dopamine levels!).

Each morning, aim to sing for five minutes before work, whether this is in your home, your car, or on your walk. In addition, utilize music, as I did, during a moment of stress. Observe what this release of endorphins does to calm your mind and body down.

If the opportunity ever presents itself, dance and sing with others. It could be a concert, in a bar, at karaoke, or at your home. Singing and dancing together has been shown to provide incredible benefits to the quality of your relationships.[34] This enables you to experience an oxytocin release and endorphin release at the same time!

Challenge

I would now like you to complete the 'Music Challenge'. In order to complete this challenge, you need to sing for five minutes every morning for the next seven days.

# 19

Are You Laughing Enough?

ENDORPHINS
LAUGHTER
ENDORPHINS
LAUGHTER
ENDORPHINS
LAUGHTER
ENDORPHINS
LAUGHTER
ENDORPHINS
LAUGHTER
ENDORPHINS

Let's start by simply rating how often you feel you laugh every day.

Rate yourself on a scale from 1 to 10 for how much laughter you have in your life.

1 → 10

1 = never
10 = tons

Understanding LAUGHTER

Think of a time when you really laughed. The type of laughter that almost hurts your body. Your stomach aches with pain as you let out a loud burst of sounds. Tears stream from your eyes. This is euphoria. This is endorphins. Your body is physically exerting itself and, as we now know, that provides a huge release of endorphins.[35] You may have heard the phrase 'laughter is the best medicine', and it is true.

Let's look at the science. A recent study revealed the impacts of 'laughter therapy' on individuals' stress and anxiety,[36] demonstrating just how impactful laughing is on regulating emotions. Benefits also derive from the unique aspects of laughter such as increased oxygen intake leading to reduction in stress hormones, and even increased activation of one's abdominal muscles leading to weight loss.[37] All of this ultimately leads to fantastic benefits for your physical and mental health.

Creating more **Laughter** in your life is something you can intentionally focus on in order to elevate your mood every day. Now, more than ever, we spend a huge proportion of our time behind screens and progressively less time connecting with one another. Putting yourself in the right environments where laughter can take place is our goal. Let's look at a few ways to do this.

1. YOUR WORKING LIFE

Working from home (WFH) has been hugely beneficial in many ways and there are many aspects of it that we have come to love. But one challenge I always see in companies that allow WFH is the reduction in people's daily social engagement. In an office you may grab a coffee or chat with colleagues about their weekend, and these are moments where a little bit of laughter can take place. I recommend that you get yourself into the office regularly, if you have one, for this sense of connection.

2. SOCIAL EVENTS

Laughing with others has the capacity to boost your oxytocin, too.[38] Take a moment to check in with yourself regarding your Social Life challenge from Part 2, and whether you are putting yourself in enough social environments to give you opportunities to laugh, have fun, and let go for a period of time.

3. SPEND TIME WITH KIDS

You may have noticed that kids laugh a lot. Spending time with young family members and younger friends, and immersing yourself in their world for a period of time, is a great way to create more laughter in your life. Be fully present with them, focus on what they are finding funny, be playful, and you'll find yourself laughing too.

4. HUMOROUS CONTENT

Now while the best form of laughter is in the company of others, it is not the only way to bring more humour into your life. There is plenty of hilarious content online that you could listen to and watch that can make you laugh. For example, comedy-based podcasts, funny movies, or humorous TV shows. All of these could bring a little more light-hearted joy into your day.

It is important to mention here the opposite type of content that you may be consuming – content that is worse than not funny, it is also very stressful for your mind. While rolling news is informative and at times necessary, you need to be seriously careful about how much of this content you consume. Constantly watching and clicking on negative stories will influence how you think and how you feel. Your brain is easily conditioned. If you are constantly told what is going wrong in the world, you will start to think about everything that is going wrong in your life too. Do, of course, keep yourself informed about current affairs, but don't let this content overwhelm you if you want to feel calm and happy.

Strategy

Here our focus is simple: increase the amount of time you are spending in environments that make you laugh.

1. **YOUR WORKING LIFE**
2. **SOCIAL EVENTS**
3. **TIME WITH KIDS**
4. **HUMOROUS CONTENT**

It is really important for us to link back here to the framework of endorphins. These live within your body to help you manage stress. Often when we are highly stressed, we don't prioritize social events and we end up spending more time immersed in our worrisome thoughts.

When you are stressed, you need social engagements and you need humour. It will provide your mind with the relief it needs to calm itself down and find a suitable solution to whatever is creating this stress for you. Prioritize social time.

Challenge

I would now like you to complete the '**Laughter Challenge**'. In order to complete this challenge, you need to put yourself in three environments that will get you laughing in the next seven days.

20

Get Out of Your Chair

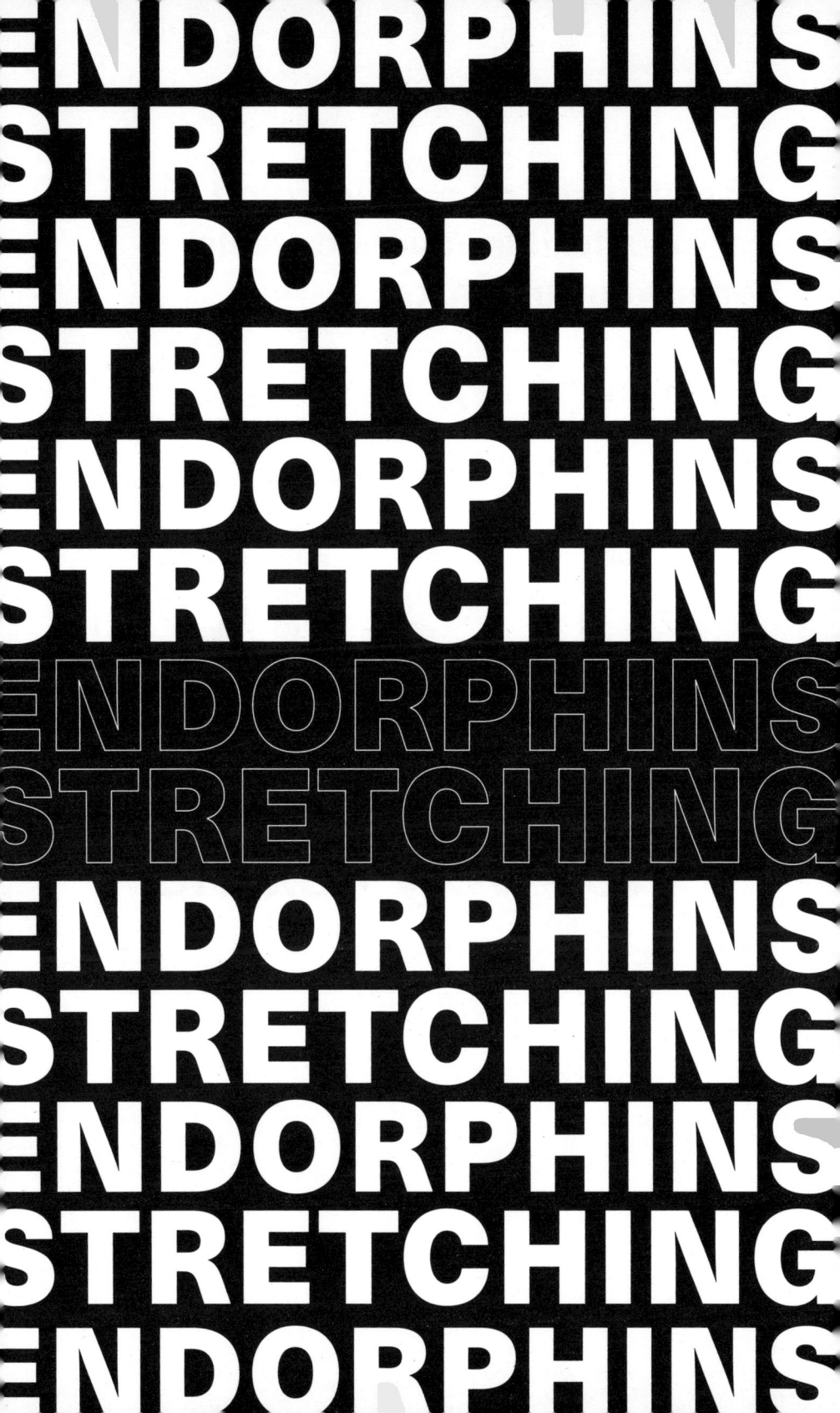

First, let's rate how mobile you are.

Rate yourself on a scale from 1 to 10 for how flexible your body feels when you stretch it.

> # 1 ➔ 10
>
> **1 = very stiff**
> **10 = very flexible**

Wow, you have made it to the final chapter of *The DOSE Effect*. Congratulations! Deciding to prioritize reading this book each day over the ease of just scrolling your phone is incredibly difficult. Take a moment to celebrate yourself. This is a huge Achievement.

Understanding STRETCHING

On to Chapter 20, Stretching. Interestingly, every time I deliver DOSE Live, this is one of the most popular sections.

Let's paint a picture of a common behaviour. You wake up, you wander around getting ready, and before you know it you're seated again. You spend many hours seated before eventually feeling hungry or needing to use the bathroom. You stand for a few minutes and very quickly you are once again seated. You spend hours, every day, in this position. Picture your future self. If you are fortunate enough to live into your old age, which I hope you are, I want you to imagine your life then. How is your posture, how mobile are you, how easy is it for you to get up from the sofa, to walk up the stairs, to interact with your grandkids? Not too easy, is it? **Stretching** is still not enough of a priority for most of our society. This is what leads to such huge physical challenges in our later years.

I envision an alternative future for you. One where your body has greater mobility and strength, and is mobile for many many years to come. One where you feel fit and strong in your old age. This is possible and very simple to do through just a few simple feel-good actions, every day of your life.

Remind yourself of the principle of endorphins. If we physically push our body, endorphins are released, and we de-stress our minds. A study examining the benefits of doing yoga revealed not just rises in endorphins but also a reduction in the stress hormone cortisol. By incorporating something as simple as yoga, you can introduce a double benefit – more endorphins to decrease stress, and less of your primary stress hormone.[39] This creates a euphoric feeling for people, known as 'yoga-high'. An additional study revealed similar incredible levels of activation of the endorphin systems, and this was connected to a number of additional health benefits, which included improvements in sleep, emotional regulation, and ability to navigate painful physical challenges.[40]

Our mission is to figure out how we can stretch your body a little more often. I understand some of you reading this book may already be avid yoga practitioners, stretching your bodies all the time. Those of you who are, incredible work, keep at it. The rest of you, and the camp I would put myself in, are those who can't really be bothered to stretch and find it boring. I have a plan for you.

First, I want you to check in right now with how you are sitting. I doubt you're currently standing up. Before you shuffle around or move, look at your posture. You may notice your shoulders are forward, your back, particularly your lower back, is curved, and your legs are in a right angle position. Maybe you're lying in the bath right now getting your Heat; if you are, then that's a vibe, and enjoy it. The rest of you who are seated, this position is what we want to learn to correct. We need a way to straighten and lengthen your back. To stretch the back of your legs. To get your spine twisting a little more often. When the world required much more physical labour, our bodies were constantly being moved through a wide range of movements that helped maintain their mobility.

A simple stretching routine

I am going to guide you towards a simple routine that you will do when you wake, at lunchtime, and before you go to sleep, every day, for the rest of your life. Now, that might seem daunting. The good news is that each set takes just thirty to sixty seconds.

This is going to involve three simple movements, and I want you to do this immediately after reading this paragraph.

MOVEMENT 1:
REACH-UPS

In a standing position, reach your hands as high as you can and try to touch the ceiling.

MOVEMENT 2:
REACH-DOWNS

Lean forward and reach down to touch your toes. Stretch as far as you can, until you feel it in the back of your legs. Don't hurt yourself.

MOVEMENT 3:
TWISTS

Raise your arms in front of you, with your palms facing down towards your body. Twist your arms around your body in one direction, first to the left side, then to the right side. You may hear some cracks in your back.

I WANT YOU TO DO THIS ROUTINE THREE TIMES.

SO . . .

- Reach tall, reach down to your toes
- Reach tall, reach down to your toes
- Reach tall, reach down to your toes

NOW . . .

Hands up in front of you
- Twist left, twist right
- Twist left, twist right
- Twist left, twist right

That's it. It's that simple. And it is unbelievably effective.

Start doing this three times per day and your mobility will begin to increase. Also feel the rush of blood into your brain, accompanied by a release of endorphins as you do it.

BAR HANGING

Have you ever been on a monkey bar, or a pull-up bar, and tried to hang off it? Maybe you've done this recently or maybe this is something you haven't done since you were a kid. Let's imagine now, you are walking through a park one morning and you see a seventy-five-year-old man or woman, hanging from a pull-up bar. You would be astounded, but again, I see this in your future. Simply hanging from a bar is an unbelievable way to decompress your spine, lengthen your body, and keep it as mobile and strong as possible.

Whether you go to a gym, or there is a park nearby, every time you see a bar I want you to try hanging from it. Now, of course, be careful. First place your hands on the bar keep your feet on the floor, and just slightly slacken your legs and feel what it's like to hold yourself up. As you gain confidence, take one foot off the floor, then the other. Gradually you will build up to hanging from the bar. Start with three seconds, then five, then ten. Build up to thirty seconds. Doing this on a daily basis will change your life. It will boost your endorphins in the present, and transform your quality of life in future.

Strategy

Our stretching strategy here is clear. You need to do this three times per day. When you wake, at lunch, and before bed, you are going to do your reach-ups, reach-downs, and twists. This will quickly start transforming your body's mobility.

Then, every time you have access to a bar, you need to practise your hanging. Start small and build up as you become more comfortable and confident.

Important note: If you are doing yoga, Pilates, or a longer daily practice, that is incredible and you should keep doing it. These stretches and bar-hanging movements are for anyone who needs to begin building this into their everyday life.

As you start stretching more, your body will become fitter, it will enjoy exercise more, and your capacity to get your DOSE of endorphins will rise and rise.

Challenge

I would now like you to complete the **'Stretching Challenge'**. In order to complete this challenge, you need to complete your stretching practice every day for the next seven days (and for the rest of your life, too!).

Building Your DOSE

Throughout Part 4, we have explored the true power of boosting your endorphins on a regular basis and how this will not only improve your physical health but significantly reduce your stress levels.

I now want you to take some time to consider which one of the five primary endorphin actions feels the most important for you to continue to prioritize. It would be incredible if all of these behaviours remain a priority in your life. However, selecting one core behaviour and ensuring it becomes deeply embedded in your life is essential.

WHAT WILL BE YOUR PRIMARY
Endorphins Action?

1. EXERCISE
Finding your favourite way to regularly get your body moving

2. HEAT
Immersing yourself in hot environments through saunas or baths

3. MUSIC
Singing and dancing on a daily basis, particularly in the mornings

4. LAUGHTER
Ensuring you are regularly in environments that maximise laughter

5. STRETCHING
A short, consistent routine that mobilizes your body

Be sure to chat with a friend or family member about the primary endorphin challenge you have selected!

I am so excited that you have come this far. What you have achieved throughout this book is genuinely phenomenal. I understand it is not easy to choose to read in a rare moment of free time, when there are so many other temptations to turn to. Take a second right now to say well done to yourself. You have now set yourself on an entirely new path for the rest of your life.

Our final goal is to solidify what you are going to take forward with you from this book, and select which actions feel the most impactful to your experience of life.

First we are going to build your DOSE. We are going to focus on each chemical and select which key action you will prioritize every single day. Once you are successfully smashing one action from each chemical per day, we will work to build this up further.

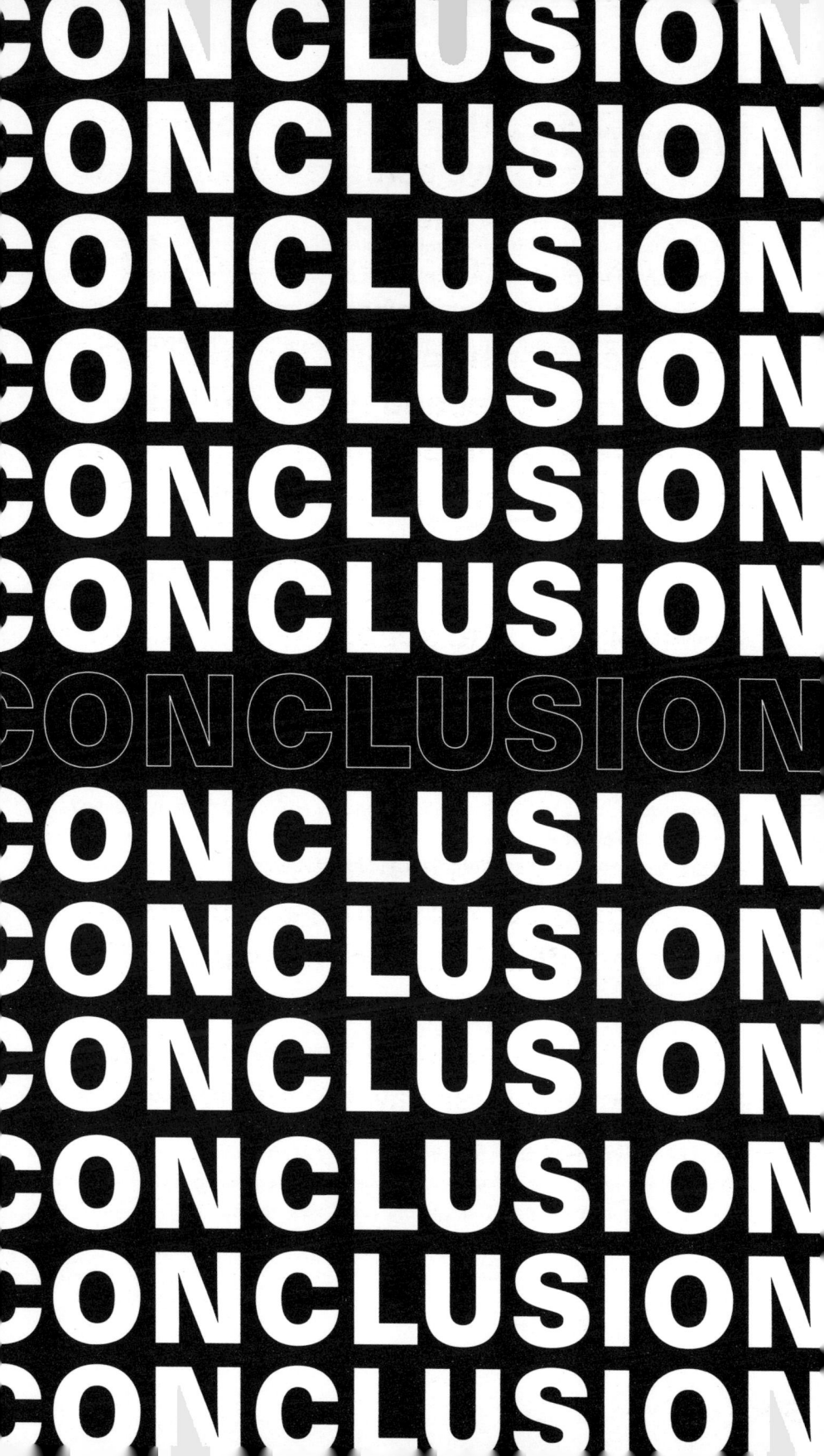

Dopamine

Let's start with dopamine. Remember, this is the chemical creating your drive in life. It controls how motivated you feel and your capacity to stay focused on your goals. It is naturally built up when you complete challenging activities. There are also unnatural spikes and crashes when you engage too frequently in the quick dopaminergic behaviours our modern world has to offer.

1. FLOW STATE

Do you feel Flow State should be the primary dopamine action you take forward? Does training your capacity to enter deep states of focus and productivity each day excite you? Can you learn to push past those first fifteen minutes of discomfort when you start a task in order to enter a blissful state of relaxed accomplishment?

2. DISCIPLINE

Does discipline need to be your primary focus right now? Intentionally building your dopamine levels by living your life in a more careful and diligent manner? Can you become disciplined at waking up and making your bed? Can you ensure your home environment remains organized and clutter free? Remember, your environment is an externalization of your mind. Clean it and observe the clarity that follows in your thoughts.

3. PHONE FASTING

Are you addicted to your phone? Be honest with yourself. I certainly am. Every day, I make a deep, conscious effort to find space from it. It isn't easy, but it is life-changing. Can you commit yourself to ignoring your phone in the morning when you first wake up? Can you commit yourself to an hour away from it each evening? Developing this capacity to find space from technology is a skill that will continue to improve the quality of your life with every year you live.

4. COLD WATER

Do you need to develop your tenacity? Your willpower? And your ability to embrace discomfort? Do you want to feel strong and empowered when facing challenges? Cold water immersion provides you with the training to do precisely this. It comes as no surprise that it hurts both psychologically and physically when the cold water hits your body. But the feeling it creates, of accomplishment, energy, and drive, is well worth it.

5. MY PURSUIT

Do you need a clear goal in your life? A goal so meaningful that you are willing to sacrifice moments of indulgence to get it? Can you define a goal that feels so inspiring to you that in order to arrive at it you let go of constantly checking your phone, eating sugar, drinking alcohol, and watching porn, and do whatever is required to get there? Nothing is more powerful than a vision of a future that lights a fire inside you. Can you spend a short period of time each day outdoors, on a walk, without your phone, dreaming and planning for the future?

Select your DOPAMINE ACTION:

1. FLOW STATE
2. DISCIPLINE
3. PHONE FASTING
4. COLD WATER
5. MY PURSUIT

Oxytocin

Next, let's consider oxytocin, the magical chemical that provides a clear scientific understanding of our true purpose: love for yourself and the people around you. A chemical that is built through service: service to others, and service to yourself. Let's consider which oxytocin action you feel will have the greatest impact on your life.

6. CONTRIBUTION

Do you feel you are adequately focused on supporting and giving love to the people around you? Ask yourself whether you feel satisfied with the contribution you are making to our world. There is a powerful feeling within you that wants this, that wants to support people. Can you consciously think each day about the contribution you are making to others? Sometimes this may be big things; other times it may be simply calling someone and listening to them.

7. TOUCH

Does physically connecting with people and animals bring you peace? Does it calm your mind? Does it make you feel more connected, and more loved? Do you feel the quantity of physical connection in your life has reduced? Can you make a conscious effort to hug your friends more? Your family more? If you are in a romantic relationship, can you make the physical aspect of your relationship more prominent?

8. SOCIAL LIFE

Do you like to socialize? Do you feel happy and energized when connecting with people? Can you see your friends and family more often and meet up for coffee, walks, dinners, or to exercise? Can you have a daily conscious check-in with yourself of 'have I done something fun with somebody else today?'

9. GRATITUDE

Do you want to experience an underlying feeling of happiness every day? A feeling of genuine peace? We all live in a world of more. More money, more clothes, better houses, better holidays, more friends, more followers. We are living in a state of constant desire, but desire for what you don't have is a life of dissatisfaction. We know it's important to dream – but the absence of gratitude for what you already have is a life of unhappiness. Can you take a moment, once a day, to ask yourself, 'What is the number one thing I am grateful for in this moment?' Is it a particular person? Your home? Your job? Your health? Nature? Can you immerse your mind for a few seconds each day in the joy of what you have? A happier experience of life awaits if you choose to do so.

10. ACHIEVEMENTS

Do you want to truly believe in yourself? Believe that you can achieve what you set your mind to, whether that's a work goal, a health goal, or a friendship goal? Or simply a goal to love yourself more, to speak to yourself in a more kind and supportive way. Can you take a moment, once a day, to celebrate yourself for some small progress that you have made? This might be taking more breaks from your phone, eating healthier meals, concentrating better when working, or talking to yourself in a more uplifting way when looking in the mirror. A simple daily action of positively reinforcing the changes you are making to your life will transform your inner voice and lead you towards a healthier experience of life.

Select your OXYTOCIN ACTION:

- ⑥ CONTRIBUTION
- ⑦ TOUCH
- ⑧ SOCIAL LIFE
- ⑨ GRATITUDE
- ⑩ ACHIEVEMENTS

Serotonin

Let's now consider serotonin. The natural chemical. The chemical that simply wants us to be a human. Someone who embraces the outdoors, eats natural foods, sleeps deeply at night, and breathes in a way that creates peace in your heart. Serotonin is here to guide you towards the behaviours that will optimize your mood and energy. It is predominantly built within your gut. Let's figure out which serotonin action will have the greatest impact on your life in the future.

11. NATURE

Have you begun to feel the power of the natural world? Connecting with your instincts is the key to unlocking the ability to make healthier decisions. Being able to feel what your body truly wants is a skill you must develop. Spending a short period of time in the natural world each day is vital. A simple walk in a park, a forest, on the beach, or by a river is how you can achieve this. When you are out there, can you disconnect from technology and reconnect with the world around you? Can you listen, look, smell, and touch the nature around you?

12. SUNLIGHT

Does sunlight have a positive impact on how you feel? Have you experimented with seeing sunlight before social media each morning? Could making a conscious effort to go outside more often be your method of raising your serotonin levels? As you know, this can be in nature, but it doesn't have to be. A moment outside a coffee shop, or sitting in the fresh air at home, provides the opportunity you need to increase your serotonin and create a happier, more energetic brain.

13. GUT HEALTH

Does the answer to boosting your serotonin lie within the decisions you make about what you eat and drink? Does a life in which your body is fuelled with natural, nutritious foods sound appealing? When energy crashes disappear and a calm, consistent, positive mood arises in their place? Can you let go of ultra-processed foods and opt for whole foods instead? Can you raise your protein intake? Increase your fruits and vegetables? And ensure your body is adequately hydrated every day?

14. UNDERTHINKING

Does your mind feel busy? Do you overthink? Do you get worried regularly? Can you make a conscious effort to learn how to make your mind and body slow down? This is within your power. You can decide to start a simple, slow breathing practice for just a few minutes each day. A breathing practice that not only trains your capacity to navigate the challenging moments but ultimately results in them hardly ever happening. Can you respond to overthinking through breathing in a certain way, alongside either vocalizing your fearful thoughts to someone you trust or returning your mind to a state of gratitude, so it can feel safe once again?

15. DEEP SLEEP

Does a proper night's sleep change how you feel the next day? Have you experienced in your DOSE journey waking up and feeling immediately ready for your day ahead? This is only possible if you get the phone out of your hand and you get yourself to sleep at a reasonable hour. Your body needs sleep. Your brain does too. Can you make sleep a greater priority in your life? Are you willing to sacrifice late nights in favour of earlier, happier mornings?

Select your SEROTONIN ACTION:

- 11   NATURE
- 12   SUNLIGHT
- 13   GUT HEALTH
- 14   UNDERTHINKING
- 15   DEEP SLEEP

Endorphins

And finally, endorphins. The chemicals that are a gift to us. The chemicals that have the capacity to de-stress your brain at any moment. Our modern world is fast and, if we are honest, it's stressful. The fact we have been given a chemical that can immediately reduce our stress levels is incredible. Ensuring that you activate endorphins each day is essential if you want to experience a relaxed and peaceful life.

16. EXERCISE

Do you want to feel fitter? Do you want to feel stronger? If you ask yourself at this moment, 'Do I need to exercise more?', what answer would you hear from your brain and body? If the answer is yes, then this is your endorphins in action. Your goal with exercise is sustainability. It's not to start some big exercise regime, only to give it up two weeks later. Your goal is to explore what form of exercise you can do. Remember, exercise isn't something you don't necessarily enjoy at first. It is challenging, but it is something that over time your body will learn to enjoy. It is incredibly important you find your chosen way to incorporate it into your day.

17. HEAT

Does heat make you feel calm? Have you got in a bath or sat in a sauna and felt calmer as a result? Could a simple daily practice of immersing yourself in a hot environment, away from your phone, be your method of boosting your endorphins?

18. MUSIC

How does music make you feel? More specifically, how do singing and dancing make you feel? Do they make you feel euphoric? Present? Happy? Can you dance more often? Can you learn the lyrics to your favourite songs and sing them more often?

19. LAUGHTER

Do you laugh enough? Are there moments when you find something so funny that every worry in your mind slips away and you are truly present? Can you increase how often you put yourself in environments where this takes place? Can you see your friends or family more often? Can you avoid complaining too much and stop talking about the doom and gloom on the news? Can you immerse yourself in the funnier side of life?

20. STRETCHING

Does your body feel like it needs stretching? If you ask your body right now, 'Would you like me to stretch you more?', what instinctive feelings arise? Is the answer yes? Can you wake up each morning and do your reach-ups, reach-downs, and twists? Can you hang from a monkey bar more often to stretch your back and shoulders? Can you take up yoga?

Select your ENDORPHINS ACTION:

- 16 EXERCISE
- 17 HEAT
- 18 MUSIC
- 19 LAUGHTER
- 20 STRETCHING

Your DOSE Actions

Now you have chosen your daily DOSE Actions, we are going to consider how you move through your day, and which actions feel the most important to you for this. On the next few pages, highlight one action per chemical that aligns to your goals. You can take a photo, write it down in your notes app, or make a copy and print it out.

DOSE Mornings

MORNING DOPAMINE ACTION:

- FLOW STATE
- DISCIPLINE
- PHONE FASTING
- COLD WATER
- MY PURSUIT

MORNING OXYTOCIN ACTION:

- CONTRIBUTION
- TOUCH
- SOCIAL LIFE
- GRATITUDE
- ACHIEVEMENTS

MORNING SEROTONIN ACTION:

- NATURE
- SUNLIGHT
- GUT HEALTH
- UNDER-THINKING
- DEEP SLEEP

MORNING ENDORPHINS ACTION:

- EXERCISE
- HEAT
- MUSIC
- LAUGHTER
- STRETCHING

DOSE Evenings

EVENING DOPAMINE ACTION:

- FLOW STATE
- DISCIPLINE
- PHONE FASTING
- COLD WATER
- MY PURSUIT

EVENING OXYTOCIN ACTION:

- CONTRIBUTION
- TOUCH
- SOCIAL LIFE
- GRATITUDE
- ACHIEVEMENTS

EVENING SEROTONIN ACTION:

- NATURE
- SUNLIGHT
- GUT HEALTH
- UNDER-THINKING
- DEEP SLEEP

EVENING ENDORPHINS ACTION:

- EXERCISE
- HEAT
- MUSIC
- LAUGHTER
- STRETCHING

DOSE Work Day

WORK DOPAMINE ACTION:

- FLOW STATE
- DISCIPLINE
- PHONE FASTING
- COLD WATER
- MY PURSUIT

WORK OXYTOCIN ACTION:

- CONTRIBUTION
- TOUCH
- SOCIAL LIFE
- GRATITUDE
- ACHIEVEMENTS

WORK SEROTONIN ACTION:

- NATURE
- SUNLIGHT
- GUT HEALTH
- UNDER-THINKING
- DEEP SLEEP

WORK ENDORPHINS ACTION:

- EXERCISE
- HEAT
- MUSIC
- LAUGHTER
- STRETCHING

DOSE Rest Day

REST DOPAMINE ACTION:

FLOW STATE · DISCIPLINE · PHONE FASTING · COLD WATER · MY PURSUIT

REST OXYTOCIN ACTION:

CONTRIBUTION · TOUCH · SOCIAL LIFE · GRATITUDE · ACHIEVEMENTS

REST SEROTONIN ACTION:

NATURE · SUNLIGHT · GUT HEALTH · UNDER-THINKING · DEEP SLEEP

REST ENDORPHINS ACTION:

EXERCISE · HEAT · MUSIC · LAUGHTER · STRETCHING

DOSE Stacking

Welcome to DOSE Stacking, which is a simple, engaging, and fun activity that really takes your knowledge and capacity to the next level.

Your goal is to think of an activity you can do once a week that enables you to achieve multiple DOSE Actions at once, thereby stimulating different brain chemicals simultaneously! Let's consider some examples.

1. A SOCIAL HIKE
Spend some time in nature, outside in the sunlight, away from your phone, while socializing with friends. During your time out there, you hike and push your body, eat some healthy food, and listen to and sing some music.

During this activity, you will activate all four chemicals through Phone Fasting, Social Life, Nature, Sunlight, Gut Health, Exercise, and Music.

2. A DINNER PARTY
Invite friends or family round for a healthy dinner. While you are chatting, consider discussing something you are grateful for at the moment and a recent achievement. Be sure to put all your phones in another room throughout the evening and have a good laugh.

During this activity, you will activate all four chemicals through Phone Fasting, Social Life, Gratitude, Achievements, Gut Health, and Laughter.

DOSE Stacking Challenge
Put your own DOSE Actions together and give yourself a massive boost!

3. A FRIENDLY SPORTS EVENT

Create a plan to play a sports game of your choice in a local park. Invite some friends and make sure you all contribute to the day. Give one another a hug when you arrive. Play some music. Bring some healthy snacks. And while playing the sport, aim to enter your Flow State!

During this activity, you will activate all four chemicals through Flow State, Exercise, Contribution, Touch, Social Life, Nature, Sunlight, Gut Health, and Music.

4. A PRODUCTIVE WORK SESSION

Create a plan for your day to get a period of deeply focused work done. Ensure you wake up and have a disciplined morning, walking in the sunlight before you see social media. Once you are ready to get in the zone, make yourself a healthy drink such as herbal tea or coffee, put your phone in another room, and put some lo-fi music on. No matter how bored you get or how challenging the task is, do not let yourself get distracted for forty-five minutes. Once you have completed the activity, celebrate the achievement and spend some time chatting with a friend on a video call.

During this activity, you will activate all four chemicals through Flow State, Discipline, Phone Fasting, My Pursuit, Social Life, Achievements, Sunlight, Gut Health, Exercise, and Music.

5. A DEEP CLEAN

Take an hour to really deep clean a room in your home. This is a great contribution to make to yourself and, if you live with people, others too. Put some music on. Chuck your phone in another room. Ensure you have a clear goal of precisely what you want to achieve for My Pursuit. Get yourself in the zone and then celebrate the achievement by watching a movie without your phone in the evening and then having an early night.

During this activity, you will activate all four chemicals through Discipline, Phone Fasting, My Pursuit, Contribution, Achievements, Deep Sleep, and Music.

The DOSE Revolution

Congratulations, you have now experienced The DOSE Effect! This is a huge achievement. I hope you feel immensely proud of your commitment to reaching this part of your DOSE journey, and I hope you are deeply experiencing the benefits in your life as a result.

The path outlined in this book is one you will continue to walk for the rest of your life. As you now so clearly understand through your recently developed neuroscience insights, our brains and bodies are something we must consistently work on and support. Bring DOSE into your life, your relationships, and your conversations. The more deeply your mind understands the value of the DOSE Actions, the more effortlessly they will continue to become an intricate part of how you live.

There are, of course, going to be moments when your new healthy habits slip, or the temptation of quick dopamine takes hold, and that's okay. That's part of living in the twenty-first century. At these moments, always remember to check in with how you are feeling, and specifically connect with how each of your decisions is impacting your motivation, relationships, energy levels, and mood. The more you can feel how the positive and negative behaviours in your life are affecting you, the smarter your daily decisions will become.

If you happen to step off the DOSE path, then simply pick up this book again and journey back in. You could start from the beginning, you could pick a specific chapter or action that feels pertinent, or you could head to Your Daily DOSE and create a new plan. There is also an audio version of *The DOSE Effect*, read by me, with some really cool additional tips and tricks. It's perfect for when you are getting ready in the morning, driving your car, or commuting to and from work. Listening to this could reignite your love for DOSE and ultimately your love for yourself.

If you feel *The DOSE Effect* has positively impacted your life, I'd love to hear about it. I read all my messages and the most fulfilling part of my existence is when I hear these actions have elevated your lives. If you feel inspired to share these ideas with your friends and family, please do. We are starting a DOSE revolution and we need each of you to play your part in guiding the people of our world in a new direction.

Thank you for walking this path with me. This is just the beginning.
I'll see you soon.

Index

A
accomplishment, feelings of 58, 59
accountability 47
achievements 113, 153–9, 161, 177, 287
addiction: dopamine spikes 31–3
   managing through goals 89–90
   phone addiction 62–75, 284
   willpower and 35
ADHD (attention deficit hyperactivity disorder) 50–2
adrenaline 80, 82
alcohol 27, 31, 85, 129, 201, 221
ancestors 146
   circadian rhythm 184
   dopamine 23, 89, 168
   endorphins 228, 231
   Flow State levels 43
   heart rate 208
   oxytocin 10, 13, 102, 168
   pleasure-pain balance 25
   serotonin 10, 13, 168
anger 230
animal shelter volunteers 114
anterior mid-cingulate cortex (AMCC) 35–6, 53
anticipatory dopamine 73
anxiety 18, 138, 169, 201, 220, 260, 267
appearance 106, 138
artistic activities 45, 52
attention 42, 134–5, 136, 139, 217
Attention Restoration Theory (ART) 177

B
babies, oxytocin and 100, 119
bar hanging 278
baths 251–3, 255
bedrooms 58, 61, 219
beds, making 60, 61, 69
bedtimes 150, 222
blood donation 113
Blue Zones 192, 195, 199
bodyweight training 242
bonding 100
books 33, 45, 220
the brain: brain-body connection 166–8
   endorphins as natural de-stressor 229
   neuroplasticity 156
   pleasure-pain balance 23–5, 79–80
   willpower and 35
breathing strategies 208, 209–11, 213, 222, 251

C
caffeine 199–200, 221
calming practice 211, 213, 222, 261
carbohydrates 195
cardio exercise 244
career goals 91, 94
charitable work 113–14
childcare 112
chores, household 60
circadian rhythm 184, 218
cleaning 45, 58, 59, 60, 61, 112, 297
cocaine 80
coffee 199–200
cognitive ability 254
cold water immersion 69, 70, 77–85, 97, 285
community 114
companionship 102
comparison 104, 145
competition 246
compliments 135
concentration 18, 40–53, 97, 197
confidence 103, 138–40
connections: familial 91
   physical 116–25, 161
   social 33, 72, 101, 104, 126–41, 161, 232, 269, 286
   oxytocin and 100, 106
contribution 109–15, 139, 161, 286
cooking 112
cooperation 111
cortisol 119, 185, 200, 231, 260, 273
cranial nerves 166
creativity goals 92–3, 94
Csikszentmihalyi, Mihaly 43
cycling 244–5

D
demotivation 27, 65
depression 82, 220
detoxification 254
diet 169, 192–8, 221
digital-based lives, impact of 101, 104, 114, 170, 177
dinner parties 296
discipline 54–61, 69–70, 97, 284
discomfort 76–85, 285
dishes, washing 61
dispositional gratitude 148
distractions 47, 175
donating possessions 114

dopamine 9, 21–97, 157, 200, 284–5, 293, 294, 295
   cold water immersion 77–85, 285
   discipline at home 54–61, 284
   exercise and 244–5
   Flow State 41–53, 284
   goals 86–95, 285
   high levels 34–5, 39
   low levels 27–33, 34, 39
   and our ancestors 23, 89, 168
   phone fasting 62–75, 284
drinks 133, 199–201
drugs 27, 31
dynorphin 26

E

eating habits 71–2, 192, 193, 194
education 52, 112
80:20 rule 196–7
elderly care 114
emergency calls 220
emotions 112, 122, 165–6, 217, 261
empathy 111
endorphins 9, 10, 13, 225–81, 290–1, 292, 293, 294, 295
   exercise and 229, 234–47, 281, 290
   heat 248–55, 281, 290
   high levels 232, 233
   laughter 231, 264–9, 281, 291
   low levels 230–2, 233
   and our ancestors 228, 231
   singing 256–63, 281
   stretching 271–9, 281, 291
endurance training 244–5, 247
energy drinks 199
energy levels 165, 168, 169, 184, 192–3, 197, 260
environment 114, 219
euphoria 267
evening routines 71–3, 220
   achievements practice 157
   DOSE evenings 293
   evening sunlight 186, 187
   gratitude practice 150
excitement 34
exclusion, feeling of 102
exercise 33, 45, 51, 71, 131, 231
   and endorphins 229, 234–47, 281, 290
   motivation and 18, 246, 247
expectations 106, 146
extroversion 129
eye contact 136, 139

F

face washing 69
family 91, 94, 112–13, 149, 179
fasting, intermittent (IF) 197–8
fatigue 177, 201
fermented foods 201
fibre 195
finances 112, 149
fitness see exercise
flexibility 272
Flow State 41–53, 59, 60, 97, 221, 284
focus 18, 43–53, 68, 97, 261
food banks 113
forest bathing 178
friends 112–13, 149, 179
fruit 193, 203
frustration 230
fundraising 114

G

gambling 27, 31
goals 20–97, 177, 246, 285
gratitude 142–51, 157, 161, 177, 212, 213, 222, 287
group support 111
the gut: brain-body connection 166–8
   gut feelings 165–6
   gut health 188–203, 224, 288
gyms 45, 245

H

habenula 220
hangovers 85
happiness 33, 105–6, 129–30, 146
hard work, effect of dopamine on 23
Harvard Study of Adult Development 129–30
heart rate 207–8, 222, 244, 260
heat 248–55, 281, 290
herbal tea 199, 221
Hof, Wim 79
home, building discipline at 33, 54–61, 97
hormones 198
household chores 60
Huberman, Andrew 32–3
hugs 119, 120–1, 122, 123, 124, 125, 136
humorous content 268
hunger 194
hydration 199–201
hyperfocus 52
hypothalamus 23

I

immune system 178, 183, 254
impulsiveness 50
inattentiveness 50
inner critic 104, 145, 156
intermittent fasting (IF) 197–8
introversion 102, 129
irritability 217
isolation, social 102, 103

J

Japanese Society of Forest Therapy 178
journalling 45, 212, 222

K
karoshi 178
kids, spending time with 268
kimchi 201
kindness, random acts of 115
kombucha 201, 221

L
lactobacillus 200
laughter 231, 264–9, 281, 291
laziness 60
learning 149, 217
legumes 195–6
Lembke, Anna 23
lethargy 60, 65, 66
Li, Dr Qing 178
lifestyle, sedentary 13, 231, 246
light: artificial 184, 219
   sunlight 51, 180–7, 288
listening 134–5, 136
loneliness 102, 103, 114, 123, 129, 141
love 102, 106
lower body exercises 240–1, 243
lunchtime sunlight 186, 187

M
McIntyre, Michael 79
mammalian dive response 84
martial arts 245
massage 124
mealtimes 133, 296
melatonin 186
memory 217
mental health 131, 164, 165, 166, 177
mentorship programmes 114
metabolism 217
the mind 165, 222
mindset, positive 177
mood 168, 183, 184, 209, 217
   low mood 169, 201
   serotonin and 165, 192
morning routines 51, 69–70, 157
   achievements practice 157
   daily calming practice 211, 213
   DOSE mornings 292
   gratitude practice 150
   morning sunlight 185, 187, 218
motivation 18, 97, 197
   dopamine 22, 34, 246, 284
   exercise and 18, 246, 247
movement 33, 218–19
muscles 238, 254
music 84, 132, 222, 256–63, 281, 290

N
National Institute on Aging 102
natural killer cells 178
nature 131, 149, 175–9, 224, 288
negativity 101, 156, 207
nerves 166, 207
nervous system 165, 207
nervousness 169
neuroplasticity 156
notifications 73–4
nutrition 18, 33, 149, 188–203, 224

O
opportunities 149
organization 112
overeating 192–3
overthinking 44, 207–12, 213
oxytocin 9, 99–161, 245, 263, 286–7, 292, 293, 294, 295
   contribution 109–15, 161, 286
   gratitude 142–51, 161, 187
   high levels 106, 107
   low levels 103–6, 107
   and our ancestors 10, 13, 102, 168
   physical connections 116–25, 161, 286
   self-belief 152–9, 100, 161
   social connection 126–41, 161, 286

P
pain-pleasure balance 23–6, 79–80
parasympathetic nervous system 208
Pelz, Dr Mindy 198
pets 121, 124
phones 104, 220
   healthy use of 18, 132
   phone fasting 62–75, 97, 232, 284
   removing phones in social settings 133, 134
   scrolling 14, 30
physical connections 116–25, 136, 161
physiological sigh breathing 210
phytoncides 178
pleasurable behaviours 31
pleasure-pain balance 23–6, 79–80
pornography 27, 31
portion control 194
positivity 101, 155–6, 158, 159, 177, 183
possessions, donating 114
posture 139, 273, 274
pro-social behaviours 111
probiotics 200–1
procrastination 23, 27
productivity 34, 47
protein 194–5, 203
pursuits 86–95, 97, 177

Q
quality time 112
quiet time 176

R
reach-downs 276
reach-ups 275
reading 33, 45, 69, 220

relationships 100, 122, 129–30, 131, 147
relaxation 131, 248–55, 281, 290
resistance bands 241
resonance breathing 209, 222
rest days, DOSE 295
reward feeling 23
romantic relationships 121, 122
running 45, 229, 244

S

sad music 261
Sandstrom, Gillian 114
satiety 192, 194, 195
sauerkraut 201
saunas 251, 253–4
screens 73, 220, 231, 267
Seasonal Affective Disorder (SAD) 183
sedentary lifestyles 231, 246
self-belief 18, 100, 103, 152–9, 161
self-care 124
self-confidence 139
self-control 57
self-criticism 106
self-talk, positive 18, 101, 155–6, 158, 159
selflessness 111–12
serotonin 9, 163–224, 245, 288–9, 292, 293, 295
  gut health 188–203, 224, 288
  high levels 170, 171
  low levels 169–70, 171
  and our ancestors 10, 13, 168
  rediscovering nature 172–9
  sunlight 180–7, 224, 288
  underthinking 205–13, 289
sex 122
Shirin-Yoku 178
shopping online 27, 31
showers 69, 70, 77–85, 253
sigh breathing 210
silence, time spent in 94–5
singing 256–63, 281
sleep 18, 33, 169, 183, 186, 214–23, 224, 254, 261, 289
snacks 193, 203
social confidence 138–40
social life and connections 33, 72, 101, 104, 179, 268, 286, 296
  prioritizing 126–41, 161, 232, 269
social media 51, 74, 220
  addiction to 30, 65, 72
  and dopamine 27, 29–30, 65, 66, 72
  inner critic 104, 145
  phone fasting 70, 71
  Social Media Moments (SMMs) 74–5
sport 52, 71, 245, 297
Spotlight Effect 138, 139
standing tall 139
steam rooms 251, 253–4
strangers, connection with 114

strength training 238–43, 247
stress 18, 228, 229, 230, 231, 251
  reducing 119, 121, 259–60, 267
stretching 271–9, 281
sugar 15, 27, 31, 51, 169, 193, 221
sunlight 170, 180–7, 218, 224, 288
support 112
surprises 113
swimming 244–5
sympathetic nervous system 207–8

T

tasks 46–7
tea 199
technology, night-time 220
teeth brushing 69
television 72, 73, 220, 268
temperature, room 219
thoughts, underthinking 44, 205–13
TikTok 66
timer challenge 48
touch 117–25, 136, 161, 286
tryptophan 193
twists 276

U

ultra-processed foods (UPFs) 191, 196–7
underthinking 44, 205–13, 222, 289
upper body exercises 238–9, 242

V

vagal tone 208, 209
vagus nerve 166, 207, 208
vegetables 195
volunteering 113

W

Waldinger, Dr Robert 129
walking 131, 150, 157, 175–9, 224, 244, 296
washing dishes 61
water 199
weight lifting 238–41
whole foods 196
willpower 35–6, 75
winter months, sunlight in 187
work 113–14, 267, 294, 297
workspaces 59, 61
worries 44, 207
writing 45

Y

yoga 273
yoghurt 201

Acknowledgments

Dad
To my inspiring dad, Thomas Power (the real Thomas Power). Thank you for encouraging me to dream big. You told me every single day when I was young, 'We become what we think about.' You taught me that no mountain is too big to climb.

Mum
To my loving mum, Penny Power. Thank you for solving every problem I've ever come to you with. Thank you for cleverly teaching me to love the feeling of being disciplined. Thank you for guiding me to be loving, to feel others' emotions, and to believe in myself.

Hannah Power
To my beautiful sister, Hannah Power. Thank you for being right by me throughout this writing journey, for being my fellow dopamine addict, and for inspiring me to take care of my health, manage my addictions, and live my life in flow.

Ross Power
To my legend of a brother, Ross Power. Thank you for teaching me to understand that hard work is the key to life. Watching you strive for greatness during my teenage years and beyond inspired me to follow your path. Thank you for opening my eyes to the idea that this path was even possible, my bro.

Georgia Farrar
To my beautiful girlfriend, Georgia Farrar. Thank you for giving me the confidence to be fully myself. For allowing me to actually live my life as a hunter-gatherer, despite how unusual that would be to most people. Your ability to listen and guide me towards the actions that will maximise the impact of DOSE for our world is absolutely phenomenal. Thank you.

Alex Eastman
To my wise cousin, Alex Eastman. Thank you for tuning me in, inspiring my creativity, and, of course, for that evening where you truly changed my life. This book is an outcome of the immense amount of wisdom you have shared with me.

Jerry Fox
To my mindful guide, Jerry Fox. Thank you for slowing me down, helping me reflect, and for teaching me to be mindful, to hear my thoughts, and to live life at a pace that provides impact and longevity.

Jen Horner
To the efficiency and humour queen, Jen Horner. Thank you for making work and life so insanely fun while ensuring we remain hyper-effective for years to come.

Stein Kolman
To the man who made this a reality, Stein Kolman. Thank you for enabling me to find my place on the internet and for building Neurify with me. I can't wait to see what we create together.

Karam Sihra
To my second brain, Karam Sihra. Thank you for managing my hyper-precise, rather eccentric brain so amazingly. For seeing the vision that lives deep within my thoughts and for bringing that vision into our world.

Clay Jubran
To the genius of online education, Clay Jubran. Thank you for helping me discover the smartest way to share DOSE with the online world. Your capacity to understand our mission and bring it to our community has revolutionised its impact.

Tom, Bev, and the team at HarperCollins
To my amazing managers, Tom Wright and Bev James, and the incredible team at HarperCollins. Thank you for your unbelievable level of belief in me and DOSE. From the moment we all met, I knew this team could change the world. Thank you for making that a reality.

TJ Power
And to myself...I know this isn't the norm, but given everything I said in the Achievements chapter, I feel this is appropriate to write.

Thank you for facing your fears and for stepping into the quiet, into nature, every single day, away from your phone. Thank you for listening to that critical voice in your mind, for learning to manage your addictions, and for deeply pondering the future health of our world in order to find this answer.

> **References**
>
> *All references quoted in this book are available on the DOSE website. You can access the full list of references, along with links to the original sources, by visiting* **www.thedoselab.com**.